ELEMENTARY CHEMICAL
THERMODYNAMICS

ELEMENTARY CHEMICAL THERMODYNAMICS

BRUCE H. MAHAN

DOVER PUBLICATIONS, INC.
Mineola, New York

Bibliographical Note

This Dover edition, first published in 2006, is an unabridged republication of the second printing of the work originally published by W. A. Benjamin, Inc., New York, in 1963.

International Standard Book Number: 0-486-45054-6

Manufactured in the United States of America
Dover Publications, Inc., 31 East 2nd Street, Mineola, N.Y. 11501

Preface

IN THIS BOOK I have tried to present thermodynamics in a straightforward manner, emphasizing its chemical applications and its physical interpretation while simplifying the mathematical development. The book was written because I feel that inclusion of thermodynamics in the beginning course is a feasible and important improvement in the chemistry curriculum. The most successful beginning courses have tried to teach and explain chemical phenomena in terms of a growing set of "molecular principles." It is my experience that this approach is strengthened considerably if students have learned to analyze a chemical process by thermodynamic methods. The consideration of chemical behavior in terms of energy and entropy, followed by an explanation of how the magnitude of energy and entropy changes are dictated by atomic properties, makes the descriptive chemistry taught in the beginning course an interesting and coherent subject.

All this material has been taught in a five-week period to students who had previously been exposed to stoichiometry, colligative properties, chemical equilibrium, electrochemistry, and molecular structure. Concurrent enrollment in a beginning calculus course is assumed, but the number of different mathematical operations a student must use is, I feel, remarkably small. This treatment of thermodynamics is intended not to supplant but to supplement the more rigorous development given in the physical chemistry course; accordingly, some topics have been deempha-

vii

sized. The Helmholtz free energy, which has only limited use in chemical problems, is developed in one of the exercises. The Carnot heat engine is treated as an application of thermodynamics rather than as a central (and too often dull and confusing) part of the second-law development. On the other hand, I have given considerable attention to the concept of entropy and I have emphasized the interpretation of entropy changes and chemical behavior in terms of qualitative molecular properties. The resulting gain in familiarity with the entropy concept should aid the more formal thermodynamic treatments given in later courses.

I would like to acknowledge permission to reproduce the table I have numbered 1–1 from Zemansky's "Heat and Thermodynamics," McGraw-Hill, New York, 1951; and Fig. 3–6 is redrawn from Daniels and Alberty's "Physical Chemistry," Wiley, New York, 1955. The thermodynamic data are taken from "Selected Values of Chemical Thermodynamic Properties," edited by F. D. Rossini et al., *National Bureau of Standards Circular 500.*

BRUCE H. MAHAN

Berkeley, California
October 1962

Contents

Physical Constants
and Conversion Factors

Avogadro's number N	6.0229×10^{23} molecules/mole
Faraday constant $\mathfrak{F}$	96,486 coulombs
	23,060 cal/volt equivalent
Ice point, 0°C	273.15°K
Gas constant R	0.08205 liter-atm/mole-deg
	1.9872 cal/mole-deg
1 cal	4.184 joules
	4.184×10^7 ergs
	0.04129 liter-atm

I

Introduction

THERMODYNAMICS provides the most general and efficient methods for studying and understanding complex physical phenomena, and thus it should be one of the first subjects studied by a student who has a serious interest in chemistry or any physical science. The features of the physical world that are most obvious to us are not the atomic properties, which we detect only with sophisticated devices, but rather the gross properties of matter obvious to one of our five senses or measurable with simple apparatus. These are properties such as pressure, volume, temperature, and composition—properties of matter in bulk, rather than of individual isolated molecules.

The properties of matter which are so obvious to us are called **macroscopic properties,** and they are, naturally, the first features we use to describe a physical situation. Thermodynamics deals only with these macroscopic quantities; the concept of a molecule need never be used in a thermodynamic argument.

To the student of elementary chemistry trained to speak so casually of atoms and molecules, the absence of the atomic concept from thermodynamics may seem to be a serious weakness. The opposite is true. The great strength of thermodynamics is that it makes no use of the subtle and sometimes flimsy theories of molecular structure. The only quantities and concepts which enter thermodynamics are the experimentally measurable macroscopic properties of matter. Thermodynamics provides a framework for systematizing, discussing, and relating these properties. That is one of the reasons why the subject is so important to the understanding of the physical world and why it deserves careful study by the beginning student.

Thermodynamics does more than provide a useful scientific language; it calls attention to the two macroscopic properties which are most fundamentally responsible for the behavior of matter. The first of these quantities is energy; the second, less familiar, is entropy. We say that these quantities are most fundamental because the eventual course of all physical events can be summarized by two statements:

The energy of the universe is conserved.

The entropy of the universe increases.

These are the first and second laws of thermodynamics. An understanding of these statements, and of how energy and entropy depend on the other physical properties of matter, allows us to establish criteria for predicting the extent to which a chemical reaction may proceed under a given set of conditions. That is, thermodynamics can show us how to calculate the equilibrium constant of a reaction which has never been run, just by using data obtained from measurements of the individual properties of the pure reactants and products. Furthermore, it can tell us how that equilibrium constant will vary as temperature is changed. We shall find that there is a wealth of other applications in which thermodynamic analysis is used to obtain a maximum amount of information from experiments that are conveniently done; thereby, experiments that are difficult, or practically impossible, are avoided.

Naturally, an examination of only the macroscopic properties of matter does not completely satisfy anyone familiar with the concepts and theories of molecular structure and behavior. A working scientist always tries to explain his observations of macroscopic properties in terms of what he knows of molecular structure. While thermodynamics is not a molecular theory, it does simplify the job of understanding physical phenomena in terms of molecular properties. Thermodynamics shows clearly that energy and entropy are the quantities that control the behavior of matter. Thus, theories of molecular structure have a well-defined question to answer: How do energy and entropy depend on the structure and properties of individual molecules? In what follows we shall demonstrate the use of thermodynamic concepts in solving chemical problems; but as opportunity allows, we shall also try to explain the qualitative connection between thermodynamics and molecular structure.

1-1 THERMODYNAMIC SYSTEMS

The most efficient way to learn about the behavior of matter is to conduct controlled experiments. Naturally, at any one time we can make measurements on only a small segment of the physical universe, and it is necessary to define the limits of this segment quite carefully. The part of the universe under investigation in an experiment is called the **system,** and all other objects which may act on the system are called the **surroundings.**

In the most fortunate instances, the boundaries which separate the system from its surroundings are experimentally well defined. For example, there is no question that for most reactions which occur in aqueous solution it is perfectly legitimate to claim that the glass beaker which contains the solution is not part of the reaction system, but instead forms a rather innocuous part of the surroundings. If we were concerned with the solubility of a salt in water, a few experiments with Pyrex, quartz, and porcelain containers would demonstrate that the concentration of a saturated solution is independent of the nature of the vessel in which it is contained. It would then be proper to claim that the solubility we had determined was a property of the system salt and water, and not of the system salt and water and beaker.

As another example of the importance of realizing the limits of a system, suppose we wish to investigate the relation between the pressure and volume of a certain amount of gas. If we do this experiment in a rigid steel cylinder equipped with a frictionless piston, then as we increase the pressure on the gas at a constant temperature, the change in volume is determined by the properties of the gas alone. In this case it is correct to say that the pressure-volume properties of the system are those of a gas.

On the other hand, if we had placed the same volume of gas in a rubber balloon and then varied the pressure on the balloon, the observed change in volume would be determined not only by the properties of the gas but also by the elastic properties of the balloon. In this case the system would have to be regarded as the gas and its container. Clearly, if we were interested only in the properties of the gas, we would choose to investigate it by using the steel cylinder and piston.

1–2   STATES AND STATE FUNCTIONS

After we have defined the limits of the system which we choose to discuss or investigate, we can proceed to describe the properties of the system. The purpose of this description is to allow any other scientist to reconstruct the system in all its important detail so that his replica will behave identically to the original. The description of the system therefore must be complete, but it is also desirable that it be as concise as possible. That is, it should be limited to information which actually affects the measured behavior of the system.

For example, if we have the simple system of a single particle of known mass acted upon by a known force such as gravity, experience tells us that it can be completely described by giving the values of its position and velocity. In other words, we have observed repeatedly that all single particles of mass m placed at a certain point x, y, z above the earth and given certain velocity components v_x, v_y, and v_z will behave the same. Therefore, we conclude that the only information necessary to completely describe the condition of the particle are its positional coordinates and velocity components. All other properties of the single-particle system, such as its kinetic or potential energy, are fixed by these six numbers.

The act of completely describing a system is called **specifying its state**. As we have just seen, the state of a single-particle system can be specified by only six numbers. The systems which we shall study in thermodynamics are much more complicated, however. For example, a mole of gas is a typical thermodynamic system, and since six numbers are necessary to specify the state of a single particle, it would appear that $6 \times 6 \times 10^{23}$ numbers are required to specify the state of a mole of gas. Fortunately, this much information is not necessary. In fact, for the purposes of thermodynamics the state of a mole of gas can be completely specified by giving the values of only two of the macroscopic properties pressure, volume, and temperature.

The reason for this enormous simplification lies in the nature of the common laboratory experiments with which thermodynamics is concerned. In these experiments we use devices such as meter sticks, thermometers, and manometers, or their more sophisticated equivalents, to examine the properties of systems. None of these instruments is sensitive to the behavior of single atomic particles. None of them gives us a record of the position and velocity of each atom or molecule. Instead, each of these devices is sensitive to a general property produced by the average behavior of all the particles in the system. In effect, these instruments are too large to respond to the behavior of the individual submicroscopic particles. The properties to which they do respond are macroscopic properties of the system as a whole.

As long as we use measuring devices which are sensitive only to the macroscopic properties of matter, the detailed knowledge of the position and velocity of individual atoms in a system is not required in order to describe the conditions or results of our experiments. Thermodynamics is concerned only with the macroscopic properties of matter; therefore, the thermodynamic state of a system can be completely described in terms of a few macroscopic quantities.

The macroscopic quantities that are used to specify the state of a thermodynamic system are called **state variables** or **state functions,** because their values depend only on the condition, or state, of the thermodynamic system. There are algebraic relationships between the state variables of a thermodynamic system. For instance, for a given amount of material, pressure, volume, and temperature are not all independent of each other but are connected by a mathematical relation called the **equation of**

state. The simplest example of such a relation is the ideal-gas equation of state, $PV = nRT$. Liquids, solids, and nonideal gases have equations of state which are generally considerably more complicated. The nature of the atoms in a system determines the algebraic form of the equation of state, and thus this equation is one of the properties which characterizes a system.

It is very important to understand the significance of the term "state function." We shall see presently that, in addition to pressure, volume, and temperature, energy is a state function; and there are many others. All state functions have the following important property: Once we specify the state of a system by giving the values of *a few* of the state functions, the values of *all* other state functions are fixed. We can use the ideal-gas equation of state to demonstrate this fact. When the pressure and temperature of one mole of an ideal gas are specified, the volume *must* assume the value $V = RT/P$. All other state functions such as energy also automatically assume definite values which are determined by the values of P and T.

State functions have one other important property. When the state of a system is altered, the change in any state function depends *only on the initial and final states of the system*, and not on how the change is accomplished. For example, when a gas is compressed from an initial pressure P_1 to a final pressure P_2, the change in pressure ΔP is given by the expression[1]

$$\Delta P = P_2 - P_1$$

Only the initial and final values of the pressure determine ΔP. Any intermediate values which P may have assumed in changing from P_1 to P_2 are immaterial.

This is by no means a trivial property, and it is possessed *only* by state functions. By way of analogy, we can say that any person who travels from Boston to San Francisco has changed his position (or state) on earth by an amount that depends *only* on the location of his initial state (Boston) and his final state (San Francisco). However, the distance he travels depends not only on

---

[1] The symbol Δ always means the change in a quantity, and the change is always computed by subtracting the initial value of the quantity from its final value.

the location of the cities but also on the path he takes from one to the other. Thus, position is a state function, but distance traveled is not. These are the significant properties of state functions, and it is these properties which, we shall find, make state functions so useful.

1-3 EQUILIBRIUM STATES

There are some conditions of thermodynamic systems which cannot be described in terms of state functions. For example, suppose our system is a gas confined in a cylinder with a movable piston. When the piston is motionless, as in Fig. 1-1a, the state of the gas can be specified by giving the values of its pressure and temperature. However, if the gas is suddenly compressed, as in Fig. 1-1b, its state cannot be described in terms of one pressure and temperature. While the piston is moving, the gas immediately in front of the piston is compressed and heated, whereas the gas at the far end of the cylinder is not. There is, then, no such thing as the pressure or temperature of the gas as a whole. Conditions in which the state variables are changing in time and space, called nonequilibrium states, are not treated by thermodynamics. Thermodynamics deals only with **equilibrium states** in which the state variables have values that are uniform and constant throughout the whole system.

Let us examine the criteria for an equilibrium state more carefully. First, the mechanical properties of a system must be uniform and constant. This means that there must be no unbalanced forces acting on or within a system, since any unbalanced forces would cause the volume to change continuously and we would be unable to specify the state of the system.

Second, the chemical composition of a system at equilibrium must be uniform, and there must be no net chemical reactions taking place. The occurrence of any net chemical change would inevitably change such properties of a system as its density or temperature and make specification of its state impossible.

The third and final criterion of an equilibrium state is that the temperature of the system must be uniform and must be the same as the temperature of the surroundings. Whenever a temperature

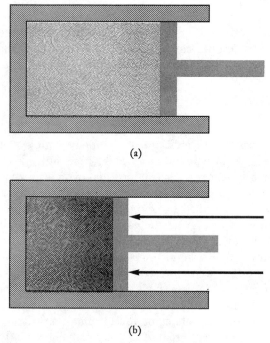

(a)

(b)

Figure 1-1 The state of a gas confined as in (a) by a motionless piston can be specified by the values of its pressure and temperature. A gas being compressed as in (b) does not have uniform properties, and its state cannot be specified by one pressure or temperature.

difference exists, heat tends to flow until the temperature difference disappears. Any system in which heat is flowing has macroscopic properties which are not uniform and which may be changing in time; therefore, such a system cannot be in an equilibrium state.

1-4 TEMPERATURE

We have emphasized that state functions are of foremost importance in thermodynamics. Some of these state functions, like pressure, volume, and chemical composition, are familiar quantities and need no elaborate explanation. On the other hand, temperature, although a common quantity, has rather subtle conceptual origins. The idea of a quantitative temperature was introduced because it became obvious that the results of many experiments depended on what we qualitatively experience as "hotness."

The first step in creating a temperature scale is to find some convenient property of matter which depends in a simple way on hotness. Temperature can be indicated, for example, by the density of liquid mercury, which is commonly measured by the distance mercury expands from a bulb into a glass capillary tube. The centigrade temperature scale is defined by assigning a value of zero temperature units, or degrees, to the length of the mercury column when the thermometer is immersed in an ice-water bath and 100 degrees to the length of the column when the thermometer is in contact with water at its normal boiling point. Intermediate temperatures are *defined* by placing 99 equally spaced marks between the two calibration points. It is clear that the temperature indicated by this thermometer depends on the properties of the materials used in its construction. Furthermore, by dividing the length between 0° and 100° into 100 *equal* units, we are really saying that temperature is something which depends *linearly* on the volume of mercury.

If some other liquid is used in the thermometer, the resulting temperature scale is different. For example, suppose we use water as the working liquid. We mark the thermometer in the same way, noting the length of the water column at the ice point and boiling point and dividing the interval into 100 equal units. Now we put our mercury thermometer and water thermometer in a bath at 0° and slowly raise the temperature. When the mercury thermometer reads 4°, the water thermometer reads −0.36°. This happens because, as the temperature is raised from 0° to 4° on the mercury scale, water contracts instead of expanding. If we used the properties of water to define temperature, we would have to say that when hotness increases, temperature sometimes goes up and sometimes goes down.

A temperature scale based on water could be used, but it would be qualitatively rather complicated. In practice, several different properties of many different materials are used to measure temperature. From our discussion we have to expect that in

Table 1-1
Comparison of Thermometers[a]

Constant-volume hydrogen thermometer $t(P)$	Constant-volume air thermometer $t(P)$	Platinum resistance thermometer $t(R)$	Thermocouple $t(emf)$	Mercury thermometer $t(l)$
0	0	0	0	0
20	20.008	20.240	20.150	20.091
40	40.001	40.360	40.297	40.111
60	59.990	60.360	60.293	60.086
80	79.987	80.240	80.147	80.041
100	100	100	100	100

[a] After M. W. Zemansky, "Heat and Thermodynamics," McGraw-Hill, New York, 1951.

general these different thermometers will read differently when they are in contact with the same body. Table 1-1 is a comparison of five types of thermometers. Even though calibrated to read the same at the boiling and freezing points of water, the thermometers deviate from one another at intermediate temperatures.

In order to allow different workers to reproduce each other's experiments, we need some reliable standard temperature scale based on an easily measured property of a readily available material. Experiments show that, at low densities, all gases at constant volume have the same dependence of pressure on temperature as measured, for example, by the mercury scale. Since all ideal gases respond identically to a given temperature change, it is convenient to redefine temperature in terms of the properties of ideal gases rather than of liquids. Temperature, then, is that quantity which depends *linearly* on the pressure of an ideal gas held at constant volume. To express this algebraically, we write

$$P = P_0 + P_0 \alpha t$$

$$t = \frac{1}{\alpha}\left(\frac{P - P_0}{P_0}\right)$$

where t is the temperature, P_0 is the gas pressure at zero degrees on the temperature scale, and $1/\alpha$ is a constant which depends on the size of the degree. When the freezing and boiling points of water (both taken at a pressure of 1 atm) are assigned the values $0°$ and $100°$, respectively, $1/\alpha = 273.15°$ and the temperature scale is called the centigrade perfect gas scale.

The ratio of two pressures P_1 and P_2, which correspond to the temperatures t_1 and t_2, is

$$\frac{P_1}{P_2} = \frac{P_0(1 + \alpha t_1)}{P_0(1 + \alpha t_2)}$$

$$\frac{P_1}{P_2} = \frac{1/\alpha + t_1}{1/\alpha + t_2}$$

This expression shows that a more convenient way to express temperature would be to define a new scale by the equation

$$T = \frac{1}{\alpha} + t = 273.15 + t$$

Then we would have

$$\frac{P_1}{P_2} = \frac{T_1}{T_2}$$

The quantity T is called the absolute, or Kelvin, temperature. From our arguments so far, it appears that this definition of temperature also depends on the properties of a specific material, the ideal gas. However, we shall see later that thermodynamics provides a way of defining temperature which is independent of the properties of any material, and the resulting scale is identical with that defined by ideal-gas properties. The fact that the Kelvin temperature is independent of the detailed structure or properties of any material is the reason it is so important in thermodynamics.

The purpose of this first chapter has been to introduce some of the language of thermodynamics. With the concepts of system, state, state function, and temperature in mind, we can begin a study of the first law of thermodynamics and its consequences.

II

The First Law
of Thermodynamics

I N THE FIRST CHAPTER we said that energy, like pressure and
temperature, is a state function. This apparently simple
statement has profound and useful consequences, as we shall soon
demonstrate. Let us start our discussion by considering the relation between energy, work, and heat.

2–1 WORK AND HEAT

One of the fundamental principles of mechanics is that the
total mechanical energy of an isolated system is constant. That
is, the sum of the kinetic and potential energies remains unchanged, even though one of the two may grow at the expense of
the other. As an example, consider a particle of mass m moving
vertically with a velocity v at a height h above the surface of the
earth. This particle has kinetic energy $\frac{1}{2}mv^2$, whereas its potential
energy is mgh. Its total mechanical energy is

$$E = \tfrac{1}{2}mv^2 + mgh$$

As the particle moves toward the earth, its potential energy decreases, but its velocity and kinetic energy so increase as to keep E constant. This behavior illustrates the general law of conservation of energy for simple mechanical systems.

If energy is conserved, any change in the total energy of one system must be compensated by an equal and opposite change in the energy of some other system. In other words, energy can be transferred but never created or destroyed. **Work** is the method by which energy is transferred from one mechanical system to another. Work is mathematically defined as the product of a force times the distance through which it acts. Thus the infinitesimal displacement dr due to external force f_{ex} produces an infinitesimal amount of work dw:

$$dw = f_{ex} \, dr$$

The work resulting from a finite displacement from r_1 to r_2 is

$$w = \int_{r_1}^{r_2} f_{ex} \, dr \qquad (2\text{--}1)$$

To show that the application of work does indeed change the energy of a system, let us treat the problem of a force acting on a simple free particle of mass m. From Newton's second law of motion we have $f_{net} = ma$, where a is the acceleration. If f_{ex} is the only force acting on the particle, we can write

$$f_{ex} = f_{net} = ma = m\frac{dv}{dt} = m\frac{dr}{dt}\frac{dv}{dr} = mv\frac{dv}{dr}$$

since $dr/dt = v$, the velocity. Now we can substitute this result for the force in Eq. (2-1) and find

$$w = \int_{r_1}^{r_2} f_{ex} \, dr = \int_{r_1}^{r_2} mv\frac{dv}{dr} \, dr = \int_{v_1}^{v_2} mv \, dv$$

$$\text{work} = w = \tfrac{1}{2}mv_2^2 - \tfrac{1}{2}mv_1^2$$

where v_1 and v_2 are the initial and final velocities of the particle. This expression says that the change in the kinetic energy of the free particle is equal to the work done on the particle by the external force.

Temperature does not enter the arguments of mechanics, since it is tacitly assumed that all parts of a mechanical system and its surroundings are at the same temperature. However, if we examine physical situations in which temperature differences do exist, we have to admit that there is another method, besides work, by which energy may be transferred. This new mode of energy transfer is called **heat,** and it operates only when temperature differences exist.

The meaning of the word "heat" is somewhat confused by an historically based tendency to think of heat as a real substance which is "contained," or which "flows." The obscurity is relieved if we realize that heat, like work, is not a material entity, but a *method* of energy transfer. The distinction between work and heat is clear; work is energy transferred by means of a mechanical link between a system and its surroundings, while heat is energy transferred solely because of a temperature difference between a system and its surroundings.

Consider now the effect of delivering a certain amount of energy, as heat, to a system. Since energy must be conserved, the heat q *added to the system* must appear as a change in the internal energy ΔE of the system, or as work w which is *done by the system*. The mathematical statement of this fact is

$$q = \Delta E + w$$

This is at once an algebraic expression of the law of conservation of energy *and of the first law of thermodynamics*. It is more commonly written in the form

$$\Delta E = q - w \qquad (2\text{-}2)$$

Equation (2–2) abbreviates the words: Any change in the internal energy of a system is the sum of the heat q which is *added to the system* minus the amount of work w *done by the system*. Notice carefully that *heat added to the system* is assigned a positive symbol q; in contrast, it is work *done by the system* which is assigned a positive symbol w. Work done on a system is denoted by $-w$. In order to prevent this inconsistency in sign convention from causing difficulty, always remember that heat added to a system, or work done on a system, must tend to raise E and thus make ΔE

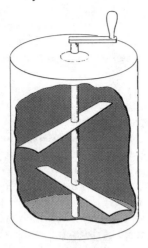

Figure 2-1 A device for raising the temperature of a liquid by performing work.

positive. Then the sign to be associated with heat or work can be deduced from Eq. (2–2).

Now we must justify our statement that E, the internal energy of a system, is a state function. To do this, we consider the problem of raising the temperature of a liter of water by $10°K$. That is, we are to take the water from an *initial state* characterized by temperature T to a *final state* of temperature $T + 10$. This change in state may be effected in a variety of ways; perhaps the simplest method would be to heat the water with a burner until the desired final state is reached. On the other hand, it is well known that we can raise the temperature of the water by performing frictional work by using a device such as that shown in Fig. 2–1.

Thus, we could obtain the desired change of state by performing work alone and allowing no heat to enter or leave the system. It is also obvious that we could use a variety of combinations of heat and work to obtain the 10° temperature rise. However, it is found *experimentally* that regardless of the path taken from the initial to the final state, that is, regardless of the individual values of q and w used, the difference $q - w$ is always the same. In other words, for a given change in state, $q - w = \Delta E$ is a constant; and since ΔE is independent of the path taken between states, *E must be a state function*.

The law of conservation of energy and the fact that E is a state function are intimately connected. To understand this, let us imagine what the consequences would be if E were not a state function. In that event ΔE would depend on the path followed between two states A and B. We could then find two ways to go from A to B: one in which we had to give to the system a small amount of energy ΔE_s and the other which required that we give a large amount of energy ΔE_l to the system to go from A to B. Then we could devise a process in which we might change the state of the system from A to B following the first path, *giving* the system energy ΔE_s, and *returning* from B to A by the second path, thereby *obtaining* energy ΔE_l from the system.

Thus, merely by taking the system from A to B and then from B to A, we would obtain an amount of energy $\Delta E_l - \Delta E_s$. There would be nothing to prevent us from repeating this cycle indefinitely, thereby creating a continuous supply of energy while leaving our system unchanged. As economically attractive as this scheme may be, all attempts to achieve it have failed. We must take this experimental demonstration of energy conservation as proof that E is a state function, that ΔE_l and ΔE_s must be equal and their difference zero.

With these ideas in mind, we can make two statements of the first law of thermodynamics, which, although superficially different, are quite equivalent. The first is

Energy is conserved when account is taken of heat effects.

The second is

For each thermodynamic system there is a quantity called the internal energy E which is a function only of the state of the system.

The first statement is just an expression of Eq. (2–2), whereas the second statement emphasizes the property of E that we shall find most useful.

Before we turn to the applications of the energy function, it is natural to ask what internal energy really is. What mechanical factors constitute internal energy? This is a question which, as far as thermodynamics is concerned, need never be answered, for thermodynamics does not offer or require any detailed interpretation of state functions such as pressure, volume, temperature, or energy. However, for those who would like to feel more secure about internal energy, we can turn to the atomic theory of matter and infer that the kinetic energy of motion of individual molecules, the potential energy due to interaction between molecules, and the kinetic and potential energy of the electrons and nuclei within individual molecules all contribute to the internal energy of a substance. When the internal energy of a substance is changed, one or more of these contributing factors changes. The value of thermodynamics comes from the fact that it shows us how to use the energy concept without demanding that we analyze it or explain energy changes on a complicated molecular basis.

2–2 PRESSURE-VOLUME WORK

There are many mechanisms by which a system may do work on its surroundings, but one of the most important is by expansion against an external force. In order to find the work associated with such a process, we start with the equation

$$w = \int_{r_1}^{r_2} f_{ex}\, dr \qquad (2\text{-}1)$$

Following the sign convention introduced in the preceding section, we say that w is the *work done by the system* on its external surroundings. The symbol f_{ex} has been used to emphasize that the force which is to be used in the computation of work is the force which links the *external surroundings* to the system. If no force links the system to its surroundings, the system can do no work.

We can write Eq. (2–1) in terms of state variables which are more useful in chemical problems by introducing the surface area A of the system:

$$w = \int_{r_1}^{r_2} f_{ex}\, dr = \int_{r_1}^{r_2} \frac{f_{ex}}{A}\, A\, dr = \int_{V_1}^{V_2} P_{ex}\, dV \qquad (2\text{-}3)$$

There are three important factors concerning this equation that must be constantly borne in mind. First, notice that work is *defined by a process*, rather than by a single state of the system. In order to integrate Eq. (2–3) and calculate the work done, *we must know how* P_{ex} *depends on* V. The relationship between P_{ex} and V depends on how a process is done, and we must expect to find that different processes yield different amounts of work even when the change in volume of the system, $V_2 - V_1$, is the same.

This point can be understood more readily by referring to Fig. 2–2. Work, as defined by Eq. (2–3), is equal to the *area* under the curve obtained by plotting P_{ex} as a function of V. Figure 2–2 shows that this area, and thus work, depends on how the system is taken from its initial state to its final state. In Chap. 1 we said that the change in any state function is independent of the path

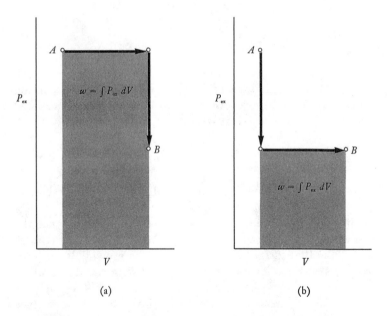

(a)   (b)

Figure 2–2 The work done depends on the path followed. The areas under the two curves shown are quite different, and thus the work done differs for the two paths even though the change of state is the same.

taken between states. Since work does not have this property, it is not a state function.

Second, we see that when a volume change occurs, the amount of work done by the system on the surroundings depends on P_{ex}, the *pressure that the surroundings exert on the system*, rather than the pressure within the system itself. If we confine a gas in a cylinder equipped with a movable piston and hold the piston at rest, then the external pressure P_{ex} we exert on the piston is equal to the pressure exerted by the gas on the piston. If, however, we violently expand or compress the gas, the external pressure exerted on the piston is different from the internal pressure of the gas. In any mechanical process that proceeds at a finite rate, forces are unbalanced, and therefore P_{ex} is not equal to the pressure of the system. In all cases we must use the external pressure to compute the work.

Finally, notice that if we follow the convention that P_{ex} in Eq. (2–3) is *always* a positive quantity, then the integral will be positive if $V_2 > V_1$. A positive value of w means that *the system does work on its surroundings*. This is what happens when a system expands, pushing back its surroundings. On the other hand, if $V_2 < V_1$, then w is negative. This corresponds to *work being done on the system*, or to a compression. To summarize, we can say that in an expansion the system does work; in a compression, work is done on the system.

We shall illustrate the calculation of work by two examples. First, suppose we have a liquid at its normal boiling point confined by a movable piston, as in Fig. 2–3a. We carry out the expansion with the external pressure on the piston equal to the vapor pressure of the liquid, which, because the liquid is at its normal boiling point, is 1 atm. Let us slowly increase the volume of the system by 1 liter. During the process, liquid evaporates and the pressure remains constant at 1 atm. Then, since $P_{ex} = P_{gas}$ is constant, Eq. (2–3) can be integrated to give

$$w = \int_{V_1}^{V_2} P_{ex}\, dV = P_{ex}(V_2 - V_1) = 1 \text{ liter} \times 1 \text{ atm}$$

The work calculated has a positive sign, which shows that during this expansion, the system *does work on its surroundings*. As we

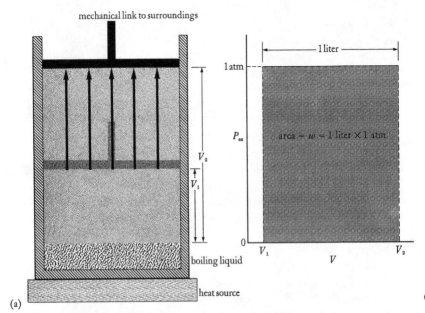

Figure 2–3 The evaporation of a liquid against a constant external pressure of 1 atm. The work done by the system is equal to the enclosed area in (b).

mentioned earlier, if we plot P_{ex} as a function of V, the work done is numerically equal to the area under the curve between the limits V_1 and V_2. Figure 2–3b shows such a plot for the process at hand. Since P_{ex} is a constant, we have only to evaluate the area of a rectangle whose dimensions are 1 liter × 1 atm. The unit of work, liter-atmosphere, is rather unfamiliar, since work or energy is usually expressed in units of ergs, joules, or calories. One liter-atmosphere of work is equivalent to 1.01×10^9 ergs, or 101 joules, or 24.2 cal.

Now let us compute the work done in a second expansion process. Again we start with the boiling liquid confined in a vessel, but this time we accomplish the expansion by allowing the vapor to pass through an opening into an initially evacuated space of 1-liter volume, as shown in Fig. 2–4. The initial and final states of the system are the same as in our preceding example,

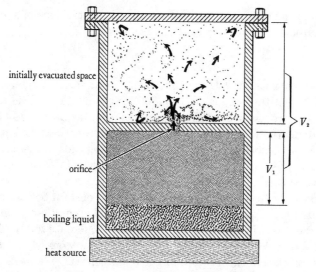

initially evacuated space

orifice

boiling liquid

heat source

V_2

V_1

Figure 2-4 The evaporation of a liquid into an evacuated space. Since there is no mechanical link to the surroundings, $P_{ex} = 0$, and no work is done.

since the volume change is the same, and again the temperature is constant. However, in this second process there is no mechanical connection between the system and its surroundings, so during the expansion, $P_{ex} = 0$. Therefore,

$$w = \int_{V_1}^{V_2} P_{ex} \, dV = P_{ex}(V_2 - V_1) = 0 \times 1 = 0.$$

No work is done, since there is no force linking the system to its surroundings. Comparison of this example with the preceding one should emphasize the fact that the work which is done in a process depends on how the process is carried out.

2-3 ENTHALPY

The fact that internal energy is a state function has immediate usefulness in the discussion of chemical processes. A chemical reaction is merely a change in the mechanical situations of a great many atoms. For example, in the reaction

$$2CO + O_2 = 2CO_2$$

the reactants CO and O_2 are one mechanical state of the atoms C and O and the product CO_2 molecules are another. The reaction is the process which takes the atoms from the initial mechanical state to the final state. Since E is a state function, the energy change ΔE associated with the reaction depends only on the nature of the initial and final states and does *not* depend on the method or path by which the reaction is run.

The ΔE of each chemical reaction run at constant temperature is a quantitative measure of the relative bond strengths and intermolecular forces in reactants and products. Reference to Eq. (2–2) and use of our expression for pressure-volume work shows how the ΔE of a chemical reaction might be measured. We have

$$\Delta E = q - w = q - \int_{V_1}^{V_2} P_{\text{ex}} \, dV$$

If we carry out a reaction keeping the volume of the system constant, that is, if

$$V_1 = V_2$$

then

$$\int_{V_1}^{V_2} P_{\text{ex}} \, dV = 0$$

and

$$\Delta E = q - 0 \qquad \text{(constant volume)}$$

$$\Delta E = q_V \qquad\qquad\qquad \textbf{(2–4)}$$

Equation (2–4) says that the change in internal energy is equal to the heat absorbed by the system *when a process occurs at constant volume*. In practice then, ΔE for a chemical reaction can be measured by running the reaction in a closed vessel, or bomb, and noting the heat evolved or absorbed as the reaction occurs.

Although q_V can be measured without too much difficulty, chemical reactions are most commonly run in open vessels, not at constant volume, but at constant pressure. Under conditions of constant pressure, the heat absorbed by the system is *not* equal to q_V or to ΔE. Therefore, despite the fact that ΔE has a rather simple physical interpretation, it is not the most appropriate quantity for characterizing chemical situations.

We shall now develop a new function of state, called the **enthalpy,** which is helpful in discussing reactions run at constant pressure. We define the enthalpy H by the equation

$$H = E + PV \qquad (2\text{-}5)$$

This definition assures us that H *is a function only of the state of a system,* since E, P, and V are all state functions. Notice that H must have the units of energy.

We can develop some of the properties of enthalpy by finding its differential dH from Eq. (2–5):

$$dH = dE + d(PV)$$
$$dH = dE + P\,dV + V\,dP \qquad (2\text{-}6)$$

This equation relates an infinitesimal change in H to similar changes in E, P, and V. To find the difference in H between two states 1 and 2, we integrate Eq. (2–6):

$$\int_{H_1}^{H_2} dH = \int_{E_1}^{E_2} dE + \int_{V_1}^{V_2} P\,dV + \int_{P_1}^{P_2} V\,dP$$

$$\Delta H = \Delta E + \int_{V_1}^{V_2} P\,dV + \int_{P_1}^{P_2} V\,dP$$

Now for ΔE let us substitute the expression

$$\Delta E = q - \int_{V_1}^{V_2} P\,dV$$

Then we get

$$\Delta H = q + \int_{P_1}^{P_2} V\,dP \qquad (2\text{-}7)$$

This is a general equation for computing ΔH for any change of state. Suppose, however, we have a restricted process—one done at constant pressure. Then, since $P_2 = P_1$,

$$\int_{P_1}^{P_2} V \, dP = 0 \qquad \text{(constant pressure)}$$

and

$$\Delta H = q_P \qquad\qquad \textbf{(2–8)}$$

The enthalpy change is equal to the heat absorbed only *when a process is carried out at constant pressure.* Representing the heat by q_P serves to emphasize this restriction. Because of the equality between ΔH and q_P, the enthalpy is often called the **heat content** of a system. However, this name tends to invite confusion between enthalpy and heat q, which, as Eq. (2–7) shows, are not always equal. The enthalpy change is equal to q only for processes which take place at constant pressure.

Notice carefully that for any process at constant pressure in which heat is evolved by the system, q_P, and thus ΔH, is a negative number. A negative value for ΔH means that the final state of the system has a smaller enthalpy than the initial state, since $\Delta H = H_f - H_i$. Processes for which ΔH and q_P are negative are said to be *exothermic.* On the other hand, if a system absorbs heat, q_P and ΔH are positive and the process is said to be *endothermic.*

So far, our results show that

$$\Delta E = q_V \qquad\qquad \textbf{(2–4)}$$

$$\Delta H = q_P \qquad\qquad \textbf{(2–8)}$$

In order to find the relationship between ΔE and ΔH, we return to the definition of H, Eq. (2–5),

$$H = E + PV$$
$$dH = dE + d(PV)$$
$$\Delta H = \Delta E + \Delta(PV)$$
$$\Delta H = \Delta E + P_2V_2 - P_1V_1$$

That is, ΔH and ΔE differ only by the difference in the PV products of the final and initial states. For chemical reaction at constant pressure in which *only solids and liquids* are involved, $P_2V_2 \cong P_1V_1$, and therefore ΔH and ΔE are nearly equal. However, if gases are produced or consumed in a chemical reaction, $\Delta(PV)$ may be quite appreciable and ΔH and ΔE may differ noticeably. For example, consider a reaction which involves only ideal gases and takes place at a constant temperature T:

$$aA(g) + bB(g) = cC(g) + dD(g)$$

By Dalton's law of partial pressures, PV for the products is

$$PV(\text{products}) = (c + d)RT$$

and PV for the reactants is

$$PV(\text{reactants}) = (a + b)RT$$

Therefore, $\Delta(PV)$ is given the expression

$$\Delta(PV) = [(c + d) - (a + b)]RT$$
$$\Delta(PV) = \Delta n\ RT$$

The total number of moles of gaseous products minus the number of moles of gaseous reactants is defined as Δn. In general, then, we have

$$\Delta H = \Delta E + \Delta(PV) = \Delta E + \Delta n\ RT \qquad \textbf{(2-9)}$$

In deriving Eq. (2–9) we have used the ideal-gas equation of state, $PV = nRT$. Since the quantity PV has the units of energy, so must nRT. This means that R has the units of energy per mole-degree. Naturally, the numerical value of R depends on which energy units are used, and Table 2–1 gives the value of R

Table 2–1
The Gas Constant R

$R = 0.08206$	liter-atm/mole-deg
$R = 8.314 \times 10^7$	ergs/mole-deg
$R = 8.314$	joules/mole-deg
$R = 1.986$	cal/mole-deg

for several of the possible energy units. The most useful of these units is the calorie, and it is very easy to remember that R has the approximate value of 2 cal/mole-deg. We shall illustrate the use of Eq. (2-9) by two examples.

EXAMPLE 2-1

When 1 mole of ice melts at 0°C and a constant pressure of 1 atm, 1440 cal of heat is absorbed by the system. The molar volumes of ice and water are 0.0196 and 0.0180 liters, respectively. Calculate ΔH and ΔE.

Since $\Delta H = q_P$, we have

$$\Delta H = 1440 \text{ cal}$$

To find ΔE by the expression $\Delta H = \Delta E + \Delta(PV)$, we must evaluate $\Delta(PV)$. Since $P = 1$ atm, we have

$$\Delta(PV) = P \,\Delta V = P(V_2 - V_1) = (1)(0.0180 - 0.0196)$$
$$= -1.6 \times 10^{-3} \text{ liter-atm} = -0.039 \text{ cal}$$

Since $\Delta H = 1440$ cal, the difference between ΔH and ΔE is trivial, and we can say $\Delta E = 1440$ cal.

EXAMPLE 2-2

For the decomposition of $MgCO_3$ by the reaction

$$MgCO_3(s) = MgO(s) + CO_2(g)$$

$\Delta H = 26,000$ cal at 900°K and 1 atm pressure. If the molar volume of $MgCO_3$ is 0.028 liter and that of MgO is 0.011 liter, find ΔE.

Let us divide $\Delta(PV)$ into two terms: that due to the volume change of the solids and that due to the appearance of gas:

$$\Delta(PV) = \Delta(PV)_{\text{solids}} + \Delta(PV)_{\text{gas}}$$
$$\Delta(PV)_{\text{solids}} = P(V_2 - V_1) = (1)(0.011 - 0.028)$$
$$= -0.017 \text{ liter-atm} = -0.41 \text{ cal}$$

Since 1 mole of gas appears in the products, $\Delta n = +1$, and

$$\Delta(PV)_{\text{gas}} = \Delta n\, RT = (1)(1.98)(900) = 1780 \text{ cal}$$

Clearly the change in the number of moles of gas makes the largest contribution to $\Delta(PV)$, as we mentioned earlier. Finally

$$\Delta E = \Delta H - \Delta(PV)$$
$$\Delta E = 24,200 \text{ cal}$$

2-4   THERMOCHEMISTRY

The enthalpy change associated with a chemical reaction depends only on the nature of the initial state of the reactants and the final state of the products. If, as is standard practice, the products and reactants are taken to be at the same temperature, then the ΔH of a reaction is effectively a comparison of the enthalpies of products and reactants under the same conditions. The importance of ΔH to a chemist arises from the fact that the size of the equilibrium constant of any reaction is always influenced, and is often dominated, by the relative enthalpies of reactants and products. Clearly then, the ΔH of a reaction is one of its most important characteristics.

It is possible to measure the ΔH of some reactions by direct calorimetry. For example, the reaction

$$C(\text{graphite, } 298°K) + O_2(\text{gas, } 298°K) = CO_2(\text{gas, } 298°K)$$
$$\Delta H = -94{,}052 \text{ cal}$$

can be carried out in a calorimeter with essentially complete conversion of reactants to products and without complicating side reactions. As the value of ΔH shows, 94,052 cal (94.052 kcal) of heat is evolved when the reaction proceeds at constant pressure. Since ΔH and ΔE are virtually identical for this reaction, the negative value of ΔH shows clearly that the energy of the product molecules is less than the energy of the reactants, which is one of the reasons the reaction proceeds so completely.

In contrast, there are many reactions which cannot be satisfactorily carried out in a calorimeter. The ΔH of the reaction

$$C(\text{graph}) + \tfrac{1}{2}O_2(g) = CO(g)$$

is of interest, but it is impossible to convert 1 mole of carbon completely to carbon monoxide without forming considerable amounts of carbon dioxide. The formation of this by-product is accompanied by heat evolution, and the heat contributed by the formation of carbon monoxide alone cannot be determined.

We can, however, calculate the ΔH of this reaction if we realize that the formation of carbon monoxide is just one step on

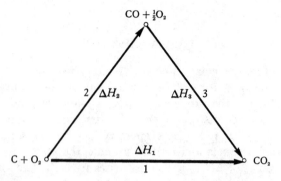

Figure 2–5 Alternate paths for the conversion of carbon and oxygen to carbon dioxide. Since enthalpy is a function of state, ΔH_1 must equal the sum of ΔH_2 and ΔH_3.

a path which leads from carbon and oxygen to carbon dioxide. Figure 2–5 represents this idea graphically. If we regard carbon plus oxygen gas as the initial state and carbon dioxide as the final state, ΔH *must be the same for any path connecting these two states.* The reason for this, of course, is that H is a state function, and its change depends only on the initial and final states and not on the path taken between them. Therefore, the ΔH of the combustion of carbon to carbon dioxide must equal the sum of the ΔH's for the combustion of carbon to carbon monoxide and of carbon monoxide to carbon dioxide.

The ΔH for the combustion of carbon monoxide can be measured easily, since carbon monoxide can be burned rapidly and completely in a calorimeter, without side reactions. The three reactions we have to deal with are

$$C(s) + O_2(g) \rightarrow CO_2(g) \qquad \Delta H_1 = -94.05 \text{ kcal}$$
$$C(s) + \tfrac{1}{2}O_2(g) \rightarrow CO(g) \qquad \Delta H_2 = \text{?}$$
$$CO(g) + \tfrac{1}{2}O_2(g) \rightarrow CO_2(g) \qquad \Delta H_3 = -67.64 \text{ kcal}$$

Since H is a function of state, we can write

$$\Delta H_1 = \Delta H_2 + \Delta H_3$$
$$\Delta H_2 = \Delta H_1 - \Delta H_3 = -94.05 - (-67.64) = -26.41 \text{ kcal}$$

A moment's consideration of the three reactions shows reaction 2 can be obtained by subtracting reaction 3 from reaction 1. In addition, we found $\Delta H_2 = \Delta H_1 - \Delta H_3$. This example illustrates an important general principle: A value of ΔH can be associated with each chemical reaction. When a number of reactions are algebraically combined, the ΔH of the resulting reaction is obtained by an identical algebraic combination of the ΔH's of the reactions.

EXAMPLE 2–3

Compute the ΔH of the reaction

$$C(\text{graph}) + H_2O(g) = H_2(g) + CO(g)$$

from the enthalpy changes of the reactions

$$C(\text{graph}) + \tfrac{1}{2}O_2(g) = CO(g) \qquad \Delta H = -26.41 \text{ kcal}$$
$$H_2(g) + \tfrac{1}{2}O_2(g) = H_2O(g) \qquad \Delta H = -57.80 \text{ kcal}$$

By subtracting the second equation from the first, we obtain the desired reaction. By combining the values of ΔH in the same way, we find

$$C(\text{graph}) + H_2O(g) = H_2(g) + CO(g)$$
$$\Delta H = -26.41 - (-57.80) = +31.39 \text{ kcal}$$

EXAMPLE 2–4

Find ΔH for the reaction

$$Mg(s) + 2HCl(g) = MgCl_2(s) + H_2(g)$$

from

$$Mg(s) + Cl_2(g) = MgCl_2(s) \qquad \Delta H = -153.2 \text{ kcal}$$
$$\tfrac{1}{2}H_2(g) + \tfrac{1}{2}Cl_2(g) = HCl(g) \qquad \Delta H = -22.06 \text{ kcal}$$

In order to obtain the desired reaction, we must multiply the second equation by 2 and subtract it from the first. The values of ΔH must be treated in exactly the same way. We find

$$Mg(s) + 2HCl(g) = MgCl_2(s) + H_2(g)$$
$$\Delta H = -153.2 - 2(-22.06) = -109.1 \text{ kcal}$$

The rule that ΔH of a reaction is independent of the number of steps or the nature of the path by which a reaction is carried out is known as Hess's law of constant heat summation. Its validity is a direct consequence of the fact that H is a state function.

2–5 ENTHALPIES OF FORMATION

In our examples we have computed the changes in H and E which occur when a system undergoes a change of state. In effect, ΔH and ΔE represent only comparisons between the properties of two states, with no absolute values of E or H specified. For the purpose of tabulation and comparison of the properties of compounds, it is extremely convenient to assign each compound an absolute enthalpy. In order to assign absolute values to H, we must make some choice of a state or condition of zero enthalpy.

The choice of this state of zero enthalpy is as completely arbitrary as the choice of having the zero of longitude pass through Greenwich, England. First, the **standard state** of each element and compound is defined as its most stable physical form at 1 atm pressure and a specified temperature, usually 298°K. By convention, *each element in its standard state is assigned an enthalpy of zero.* Then the enthalpy change that accompanies the formation of one mole of a compound in its standard state from elements in their standard states is called the **standard enthalpy of formation,** and it is given the symbol ΔH_f°. The temperature at which the formation occurs is sometimes specified, but if it is not, a value of 298°K is assumed. The subscript f in ΔH_f° shows that the reaction referred to is the formation of the compound from its elements, and the superscript ° shows that all reactants and products are in their standard states. Thus

$$C(\text{graph}) + O_2(g, 1 \text{ atm}) = CO_2(g, 1 \text{ atm})$$

$$\Delta H = \Delta H_f^\circ(CO_2) = -94.05 \text{ kcal}$$

$$H_2(g, 1 \text{ atm}) + \tfrac{1}{2}O_2(g, 1 \text{ atm}) = H_2O(g, 1 \text{ atm})$$

$$\Delta H = \Delta H_f^\circ(H_2O, g) = -57.80 \text{ kcal}$$

An enthalpy of formation can be associated with each compound, and it is a quantitative measure of the enthalpy or, approximately, the energy of a compound relative to its elements. Compounds for which ΔH_f° is negative are apt to be very stable with respect to their elements. A positive enthalpy of formation shows that a compound is formed from its elements only when

Table 2–2
Enthalpies of Formation
ΔH_f°, kcal/mole, at 298°K

Inorganic compounds

$H_2O(g)$	−57.79	$AgCl(s)$	−30.36
$H_2O(l)$	−68.32	$AgBr(s)$	−23.8
$H_2O_2(g)$	−32.53	$AgI(s)$	−14.9
$O_3(g)$	34.0	$CaO(s)$	−151.8
$HCl(g)$	−22.06	$Ca(OH)_2(s)$	−235.6
$HBr(g)$	−8.66	$CaCO_3(s)$	−288.4
$HI(g)$	6.20	$BaO(s)$	−133.5
$SO_2(g)$	−70.96	$BaCO_3(s)$	−290.8
$SO_3(g)$	-94.45	$BaSO_4(s)$	−345.3
$H_2S(g)$	−4.81	$Fe_2O_3(s)$	−196.5
$N_2O(g)$	19.49	$Al_2O_3(s)$	−399.1
$NO(g)$	21.60	$SiO_2(s)$	−209.9
$NO_2(g)$	8.09	$CuO(s)$	−37.6
$NH_3(g)$	−11.04	$Cu_2O(s)$	−40.4
$CO(g)$	−26.41	$ZnO(s)$	−83.2
$CO_2(g)$	−94.05	$ZnS(s)$	−48.5

Organic compounds
Gases

Methane, CH_4	−17.89	Ethylene, C_2H_4	12.50
Ethane, C_2H_6	−20.24	Acetylene, C_2H_2	54.19
Propane, C_3H_8	−24.82	1-Butene, C_4H_8	0.28
n-Butane, C_4H_{10}	−29.81	cis-2-Butene, C_4H_8	−1.36
Isobutane, C_4H_{10}	−31.45	$trans$-2-Butene, C_4H_8	−2.40
n-Pentane, C_5H_{12}	−35.00	Isobutene, C_4H_8	−3.34
Isopentane, C_5H_{12}	−36.92	1,3-Butadiene, C_4H_6	26.75
Neopentane, C_5H_{12}	−39.67	Methyl chloride, CH_3Cl	−19.6

Liquids

Methanol, CH_3OH	−57.02
Ethanol, C_2H_5OH	−66.35
Acetic acid, CH_3COOH	−116.4
Benzene, C_6H_6	11.72
Chloroform, $CHCl_3$	−31.5
Carbon tetrachloride, CCl_4	−33.3

Table 2–2 (*continued*)

Aqueous ions

H+	0	OH−	−54.96
Na+	−57.28	Cl−	−40.02
K+	−60.04	Br−	−28.90
Ag+	25.31	I−	−13.37
Ca²⁺	−129.77	S²⁻	10.0
Ba²⁺	−128.67	HS−	−4.22
Cu²⁺	15.39	CO₃²⁻	−161.63
Zn²⁺	−36.43	SO₄²⁻	−216.90

Gaseous atoms

H	52.1	C	171.7
O	59.1	N	112.5
Cl	29.0	Br	26.7

energy is added to the reaction system, and such compounds tend to be unstable. Table 2–2 contains the ΔH_f° of various compounds at 298°K.

Some enthalpies of formation can be measured by direct calorimetry; others must be calculated indirectly. For instance, the ΔH_f° of methane, CH_4, cannot be measured directly because the reaction

$$C(s) + 2H_2(g) \rightarrow CH_4(g)$$

cannot be carried out in a calorimeter. However, it is easy to measure the enthalpy of combustion of methane, and we can combine this with other available information to obtain $\Delta H_f^\circ(CH_4)$.

$$CH_4(g) + 2O_2(g) \rightarrow CO_2(g) + 2H_2O(g) \qquad \Delta H = -191.77 \text{ kcal}$$
$$C(s) + O_2(g) \rightarrow CO_2(g) \qquad \Delta H = -94.05 \text{ kcal}$$
$$2H_2(g) + O_2(g) \rightarrow 2H_2O(g) \qquad \Delta H = 2(-57.80) \text{ kcal}$$

If we subtract the first reaction from the *sum* of the last two reactions, we find

$$C(s) + 2H_2(g) \rightarrow CH_4(g)$$
$$\Delta H = \Delta H_f^\circ(CH_4) = 2(-57.80) - 94.05 + 191.77$$
$$\Delta H_f^\circ(CH_4) = -17.88 \text{ kcal/mole}$$

The enthalpies of formation of all hydrocarbons can be found by measuring the ΔH of combustion and then doing a calculation similar to the one above.

Besides the fact that enthalpies of formation give a convenient indication of the stability of compounds relative to their elements, they are useful in carrying out thermochemical calculations. The ΔH° of any reaction is the difference between the ΔH_f° of all products and ΔH_f° of all reactants. That is,

$$\Delta H = \Sigma \, \Delta H_f^\circ(\text{products}) - \Sigma \, \Delta H_f^\circ(\text{reactants}) \quad \textbf{(2-10)}$$

We shall demonstrate the validity of this statement with an example. The transformation

$$C_2H_4(g) + H_2O(l) \rightarrow C_2H_5OH(l)$$

can be carried out by direct reaction between the ethylene, C_2H_4, and water to give ethanol, C_2H_5OH, or, in principle, by decomposing ethylene and water to their elements and then allowing the elements to react to form ethanol. The alternate paths are indicated in Fig. 2-6. Since the enthalpy change is independent of the path, we can write

$$\Delta H_1 = \Delta H_2 + \Delta H_3$$

Step 2 is the decomposition of ethylene and water into their elements. Therefore

$$\Delta H_2 = -\Delta H_f^\circ(C_2H_4) - \Delta H_f^\circ(H_2O)$$

whereas for step 3 we have

$$\Delta H_3 = \Delta H_f^\circ(C_2H_5OH)$$

This gives us

$$\Delta H = \Delta H_f^\circ(C_2H_5OH) - \Delta H_f^\circ(C_2H_4) - \Delta H^\circ(H_2O)$$
$$\Delta \bar{H} = \Delta H_f^\circ(\text{product}) - \Sigma \, \Delta H_f^\circ(\text{reactants})$$

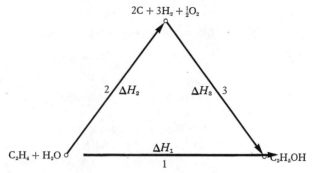

Figure 2-6 Alternate paths for the conversion of ethylene and water to ethyl alcohol.

which is consistent with Eq. (2–10). From Table 2–2 we find

$$\Delta H_f^\circ(C_2H_5OH) = -66.3 \text{ kcal}$$
$$\Delta H_f^\circ(C_2H_4) = 12.5 \text{ kcal}$$

and

$$\Delta H_f^\circ(H_2O, l) = -68.3 \text{ kcal}$$

Therefore

$$\Delta H^\circ = -66.3 + 68.3 - 12.5 = -10.5 \text{ kcal}$$

Any reaction can be conceptually dissected into two steps. In the first step, reactants are converted to their elements; the enthalpy of this process is always the negative sum of ΔH_f° for the reactants. The second step is the formation of products from the elements; the enthalpy of this reaction is always the sum of ΔH_f° for the products. Therefore, a table of enthalpies of formation allows us to calculate the ΔH° for any reaction which involves reactants and products in the table.

One of the purposes of thermodynamics is to help us use the results of experiments which are convenient, to avoid doing experiments which are difficult. As we have already emphasized, there are many reactions for which ΔH is difficult or impossible to measure. These measurements are entirely unnecessary if ΔH_f°

is known for each compound involved in the reaction. Furthermore, tabulation of ΔH_f° is an efficient way to store thermochemical information. It requires much less space to tabulate the ΔH_f° of, let us say, a hundred compounds than to list separately the ΔH of each of the many hundreds and possibly thousands of reactions which involve these substances.

EXAMPLE 2-5

Use Table 2-2 to calculate the enthalpy of combustion of ammonia to gaseous H_2O and (a) N_2, (b) NO. In the first case our reaction is

$$2NH_3(g) + \tfrac{3}{2}O_2(g) = N_2(g) + 3H_2O(g)$$
$$\Delta H^\circ = \Delta H_f^\circ(N_2) + 3\Delta H_f^\circ(H_2O, g) - 2\Delta H_f^\circ(NH_3) - \tfrac{3}{2}\Delta H_f^\circ(O_2)$$
$$= 0 + 3(-57.79) - 2(-11.04) - \tfrac{3}{2}(0)$$
$$\Delta H^\circ = -151.29 \text{ kcal}$$

In the second instance the reaction is

$$2NH_3(g) + \tfrac{5}{2}O_2(g) = 2NO(g) + 3H_2O(g)$$
$$\Delta H^\circ = 2\Delta H_f^\circ(NO) + 3\Delta H_f^\circ(H_2O, g) - 2\Delta H_f^\circ(NH_3)$$
$$\Delta H^\circ = -109.09 \text{ kcal}$$

Enthalpies of Atoms

Notice in Table 2-2 that ΔH_f° is known for chemical species whose existence in most systems is only transitory. For instance, the hydrogen atom exists in small concentrations in flames, explosions, and other reaction systems involving hydrogen-containing compounds. The ΔH_f° for the hydrogen atom is equal to one-half the enthalpy required to break the hydrogen molecule into two hydrogen atoms:

$$H_2 = 2H \qquad \Delta H^\circ = 2\Delta H_f^\circ(H)$$
$$\Delta H^\circ = 104.2 \text{ kcal}$$

Then

$$\Delta H_f^\circ(H) = \frac{104.2}{2} = 52.1 \text{ kcal}$$

The energy required to dissociate a hydrogen molecule into two hydrogen atoms can be accurately measured, and it is called the **dissociation energy** of the hydrogen molecule. The enthalpies of formation of other atoms of elements which ordinarily exist as diatomic molecules are calculated from dissociation energies by the same procedure. They can be used to find the ΔH of reactions such as

$$H(g) + Cl_2(g) = HCl(g) + Cl(g)$$
$$\Delta H° = \Delta H_f°(HCl) + \Delta H_f°(Cl) - \Delta H_f°(H)$$
$$\Delta H° = -22.1 + 29.0 - 52.1 = -45.2 \text{ kcal}$$

The enthalpy of formation of a gaseous atom of an element which is ordinarily a solid can be found by measuring $\Delta H°$ for the vaporization process. For example, the standard state of the element carbon is the graphite crystal at 1 atm and 298°K. The heat absorbed when 1 mole of graphite is vaporized to carbon atoms is commonly called the enthalpy of sublimation of carbon. Since the enthalpy of formation of graphite is zero, we can write

$$C(\text{graph}) = C(\text{gaseous atoms})$$
$$\Delta H° = \text{enthalpy of sublimation of carbon} = 171.7 \text{ kcal}$$
$$\Delta H° = \Delta H_f°(C, \text{atom}) - \Delta H_f°(C, \text{graph})$$
$$\Delta H° = \Delta H_f°(C, \text{atom}) - 0$$
$$\Delta H° = \Delta H_f°(C, \text{atom}) = 171.7 \text{ kcal}$$

That is, the enthalpy of formation of a mole of gaseous carbon atoms is equal to the enthalpy of sublimation of a mole of graphite.

Ionic Solutions

Many of the reactions which take place in aqueous solutions involve ions rather than neutral molecules. For example, when hydrochloric acid and sodium hydroxide solutions are mixed, we represent the net change by

$$H^+(aq) + OH^-(aq) = H_2O$$

rather than by

$$HCl + NaOH = NaCl + H_2O$$

Since aqueous ions are important reagents, it would be very useful if we could assign a $\Delta H_f°$ to individual ions such as H^+ and

OH^-. To see how this can be done, let us consider the formation of a 1 M solution of hydrochloric acid from hydrogen and chlorine:

$$\tfrac{1}{2}H_2(g) + \tfrac{1}{2}Cl_2(g) = HCl(g) \qquad \Delta H = \Delta H_f^\circ(HCl) = -22.1$$
$$HCl(g) + nH_2O = H^+(aq) + Cl^-(aq) \quad \Delta H = -17.9$$

where nH_2O represents the amount of water necessary to produce a 1 M solution. Addition of these two equations would give us ΔH° for the formation of H^+ *and* Cl^- in solution. Since ordinary chemical compounds are electrically neutral, ions of only one charge never can be isolated and there appears to be no way in which the ΔH_f° of an individual ion can be accurately evaluated.

Once it is accepted that only the sum of the enthalpies of formation of any pair of positive and negative ions can ever be measured, we can adopt an artificial convention to create a scale of "absolute" enthalpies for ions. If we define ΔH_f° of the proton as zero at all temperatures, then from

$$\tfrac{1}{2}H_2(g) + \tfrac{1}{2}Cl_2(g) = H^+(aq) + Cl^-(aq) \qquad \Delta H^\circ = -40.0 \text{ kcal}$$

we can write $\Delta H_f^\circ(Cl^-) = -40.0$ kcal/mole. As shown in Example 2–6, other "absolute" enthalpies can be calculated by using $\Delta H_f^\circ(H^+)$ or $\Delta H_f^\circ(Cl^-)$. These quantities are very valuable in precise calculations of the thermodynamics of ionic reactions, but it must be remembered that they are not really absolute enthalpies of formation, but are only enthalpies relative to the arbitrary value of zero for $\Delta H_f^\circ(H^+)$.

EXAMPLE 2–6

When 1 mole of a dilute strong acid is neutralized with a strong base, $\Delta H = -13.70$ kcal. The enthalpy of formation of liquid water is -68.31 kcal, and $\Delta H_f^\circ(H^+) = 0$. Use these data to find $\Delta H_f^\circ(OH^-)$. We have

$$H^+(aq) + OH^-(aq) = H_2O(l) \qquad \Delta H^\circ = -13.70 \text{ kcal}$$
$$H_2(g) + \tfrac{1}{2}O_2(g) = H_2O(l) \qquad \Delta H^\circ = -68.31 \text{ kcal}$$

Subtracting the first reaction from the second, we find

$$H_2(g) + \tfrac{1}{2}O_2(g) = H^+(aq) + OH^-(aq) \qquad \Delta H^\circ = -54.61 \text{ kcal}$$

By Eq. (2–10), ΔH° can also be expressed as

$$\Delta H^\circ = \Delta H_f^\circ(H^+) + \Delta H_f^\circ(OH^-) - \Delta H_f^\circ(H_2) - \tfrac{1}{2}\Delta H_f^\circ(O_2)$$
$$-54.61 = 0 + \Delta H_f^\circ(OH^-) - 0 - 0$$
$$\Delta H_f^\circ(OH^-) = -54.61 \text{ kcal}$$

2-6   BOND ENERGIES

Suppose we want to find the ΔH of a reaction that involves a compound whose ΔH_f° is not known. Perhaps the compound has never been synthesized or is unsuitable for use in a calorimeter. The thermodynamic techniques we have developed are useless in this situation, since thermodynamic data available are insufficient for the calculation. However, if we are willing to step beyond strictly thermodynamic considerations and use our knowledge of molecular structure, it is possible to make an estimate of ΔH.

It is found that the energy necessary to break a particular type of bond is *approximately* the same regardless of the molecule in which the bond occurs. For example, to carry out the process

$$
\begin{array}{ccc}
\quad\text{H} & \quad\text{H} & \\
\quad| & \quad| & \\
\text{H---C---H} = \text{H---C} & + \text{H} \\
\quad| & \quad| & \\
\quad\text{H} & \quad\text{H} &
\end{array}
$$

102 kcal is required per mole of bonds broken. To accomplish the corresponding change in ethane,

$$
\begin{array}{cc}
\text{H H} & \text{H H} \\
|\ \ | & |\ \ | \\
\text{H---C---C---H} = \text{H---C---C} + \text{H} \\
|\ \ | & |\ \ | \\
\text{H H} & \text{H H}
\end{array}
$$

requires approximately 96 kcal. The energy needed to break other C—H bonds generally falls near or within the range from 96 to 102 kcal. It is therefore possible to characterize the C—H bond, or any other chemical bond, by a **bond energy** ϵ which is the *approximate* energy needed to *break* that bond in any compound in which it occurs. Table 2-3 is a short list of average bond energies.

In order to use bond energies to estimate the ΔH of a reaction, we need only view the reaction as a process which breaks and makes chemical bonds. For example, in the reaction

Table 2-3
Average Bond Energies

Bond	Energy, kcal/mole	Bond	Energy, kcal/mole
H—H	103	C—H	98
C—C	80	N—H	92
C=C	145	O—H	109
C≡C	198	Cl—H	102
N—N	37	Br—H	87
N≡N	225	I—H	71
O—O	34	C—Cl	78
O=O	117	C—N	66
Cl—Cl	57	C≡N	210
Br—Br	45	C—O	79
I—I	35	C=O	173

one C=C bond and one H—H bond are broken and one C—C bond and two C—H bonds are formed. The bond energy of the bonds broken minus the energy of the bonds formed is equal to ΔE and, to an acceptable approximation, ΔH for the reaction. Thus

$$\text{bond energy in} = \epsilon_{C=C} + \epsilon_{H-H} = 145 + 104 = 249 \text{ kcal}$$
$$\text{bond energy out} = \epsilon_{C-C} + 2\epsilon_{C-H} = 80 + 196 = 276 \text{ kcal}$$
$$\Delta H° \cong \text{bond energy in} - \text{bond energy out}$$
$$\Delta H° \cong 249 - 276 = -27 \text{ kcal}$$

According to this estimate, 27 kcal should be evolved when one mole of ethylene is hydrogenated. The measured enthalpy of reaction is -32.7 kcal, which shows that the bond energy calculation is only approximate. Nevertheless, the bond energy method is useful whenever accurate values of ΔH_f° of compounds do not exist.

2–7   HEAT CAPACITIES

The amount of heat, expressed in calories, necessary to produce a standard change of one centigrade degree in one gram of material is called the **specific heat.** The product of the specific heat times the molecular weight of a substance is the heat required to raise the temperature of *one mole* of that material one degree centigrade. This is called the **molar heat capacity,** and it is a positive number which has the unit of calories per mole-degree. The molar heat capacity is more useful in chemical problems than the specific heat, and subsequently we shall use this quantity exclusively in all calculations.

Since the heat capacity C is the amount of heat needed to produce a temperature change of one degree, it would appear that the heat required to produce a temperature change ΔT is

$$q = C\,\Delta T = C(T_2 - T_1) \qquad (2\text{--}11)$$

However, there is an immediate difficulty in the application of Eq. (2–11). Heat is not a state function. Therefore, the heat involved in a process is determined not only by the initial and final states but also by how the process is done. Equation (2–11) says nothing about how the temperature is changed. We can remove this vagueness by defining two molar heat capacities; one, C_P, for processes at constant pressure, the other, C_V, for processes at constant volume. Thus, $C_P = q_P/\Delta T$, and $C_V = q_V/\Delta T$. The equivalent definitions in terms of infinitesimals are

$$C_P = \frac{dq_P}{dT} = \frac{dH}{dT} \qquad (2\text{--}12)$$

$$C_V = \frac{dq_V}{dT} = \frac{dE}{dT} \qquad (2\text{--}13)$$

From these definitions we can find that the heat required to change the temperature of n moles of material from T_1 to T_2 is

$$q_P = \int_{T_1}^{T_2} nC_P \, dT \qquad (2\text{--}14)$$

for a process at constant pressure, and

$$q_V = \int_{T_1}^{T_2} nC_V \, dT \qquad (2\text{--}15)$$

for a process at constant volume. For many substances near room temperature, C_P and C_V are constants independent of temperature. In such cases, Eqs. (2–14) and (2–15) are easily integrated to give

$$q_P = nC_P(T_2 - T_1)$$
$$q_V = nC_V(T_2 - T_1)$$

In other instances, the heat capacity depends on temperature, and the algebraic form of this dependence must be known before Eq. (2–14) or (2–15) can be integrated.[1]

For solids and liquids C_P and C_V are virtually equal, but for gases they differ noticeably. In the case of an ideal gas, the relation between C_P and C_V is remarkably simple: $C_P = C_V + R$. To derive this, we need only combine the definition of enthalpy with the definitions of C_P and C_V. Thus, for one mole of gas

$$H = E + PV$$

$$\frac{dH}{dT} = \frac{dE}{dT} + \frac{d(PV)}{dT}$$

$$C_P = C_V + \frac{d(PV)}{dT} = C_V + \frac{d(RT)}{dT}$$

$$C_P = C_V + R$$

Since most gases have a C_V which is between 3 and 10 cal/mole-deg and $R = 1.98$ cal/mole-deg, the difference between C_P and C_V is significant. Table 2–4 is a short list of the heat capacities of several substances at room temperature.

---

[1] The temperature dependence of heat capacity is often expressed by an equation of the type $C_P = a + bT + cT^2$, where a, b, and c are constants which characterize a particular substance. We are assuming that b and c are small, so that $C_P \cong a$ and is independent of temperature.

Table 2–4

Molar Heat Capacities
at Constant Pressure

H_2	6.90
O_2	7.05
N_2	6.94
CO	6.97
CO_2	8.96
CH_4	8.60
C_2H_6	12.71
NH_3	8.63
$H_2O(g)$	5.92

2–8　TEMPERATURE DEPENDENCE OF ΔH

The enthalpies of formation which are tabulated refer the amount of heat involved in the formation of the compound from its elements at a particular temperature, usually 298°K. These enthalpies of formation may be used to compute the enthalpies of reactions in which the initial and final states are also at a temperature of 298°K. It is found experimentally, however, that the enthalpy change of a reaction depends on the temperature at which the reaction is run. We would like to be able to calculate the enthalpy change for a reaction run at any temperature, rather than to have to determine this experimentally. A simple thermodynamic argument shows us how ΔH depends on temperature.

Consider Fig. 2–7, which graphically represents two possible paths for converting carbon monoxide and oxygen to carbon dioxide. All processes represented in Fig. 2–7 take place at a constant pressure of 1 atm. Step 1 is the conversion of carbon monoxide and oxygen to carbon dioxide at a temperature of 298°K. If the pressure is held constant, the heat of the reaction is ΔH_1, which we can compute from the enthalpies of formation of CO and CO_2. $\Delta H_1 = \Delta H_f^\circ(CO_2) - \Delta H_f^\circ(CO) = -67.6$ kcal. Step 2 represents the heating of the reactants at a constant pressure from

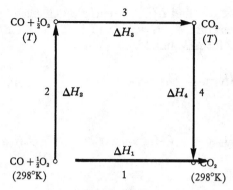

Figure 2-7 Alternate paths for converting carbon monoxide and oxygen to carbon dioxide.

298°K to an arbitrary temperature T. The enthalpy change in this process is ΔH_2, and it can be computed as

$$\Delta H_2 = \int_{298}^{T} C_P(\text{reactants}) \, dT$$

Here $C_P(\text{reactants})$ stands for the *sum* of the heat capacities of the reactants, $CO + \frac{1}{2}O_2$, at constant pressure. Step 3 is just the conversion of reactants to products at the arbitrary temperature T. The enthalpy change ΔH_3 is generally unknown, and it is, in fact, just what we want to calculate. Step 4 is a process in which CO_2 is cooled at constant pressure from T to 298°K. For the enthalpy change of this process, ΔH_4, we find

$$\Delta H_4 = \int_{T}^{298} C_P(\text{products}) \, dT$$

where $C_P(\text{products})$ is, in this example, the heat capacity of carbon dioxide at constant pressure.

Since the enthalpy change in going from $CO + \frac{1}{2}O_2$ to CO_2 is independent of the path,

$$\Delta H_1 = \Delta H_2 + \Delta H_3 + \Delta H_4 \tag{2-16}$$

If the heat capacities of the reactants and products are known, we can calculate ΔH_2 and ΔH_4, and ΔH_1 is assumed to be known.

Therefore, we can find ΔH_3, the enthalpy change of the reaction, at any temperature T. If we solve Eq. (2–16) for ΔH_3 and make substitutions for ΔH_2 and ΔH_4, we obtain

$$\Delta H_3 = \Delta H_1 - \int_{298}^{T} C_P(\text{reactants})\, dT - \int_{T}^{298} C_p(\text{products})\, dT$$

or

$$\Delta H_3 = \Delta H_1 + \int_{298}^{T} C_P(\text{products})\, dT - \int_{298}^{T} C_P(\text{reactants})\, dT$$

By defining

$$\Delta C_P \equiv C_P(\text{products}) - C_P(\text{reactants})$$

we can write

$$\Delta H_3 = \Delta H_1 + \int_{298}^{T} \Delta C_P\, dT$$

since the sum of integrals is the integral of the sum of the integrands. We see that the enthalpy of reaction depends on temperature *only through the difference in the heat capacities of reactants and products*. If reactants and products have the same heat capacities, then the enthalpy of reaction is independent of temperature. If the temperature T is not too different from 298°K, we can assume that C_P of each reactant and product is a constant independent of temperature. Then ΔC_P must also be temperature independent and

$$\Delta H_3 = \Delta H_1 + \Delta C_P(T - 298)$$

or for any two temperatures T_1 and T_2,

$$\Delta H_{T_2} = \Delta H_{T_1} + \Delta C_P(T_2 - T_1) \tag{2-17}$$

Now let us find ΔH_{398}° for the carbon monoxide–oxygen reaction. From Table 2–4 we find $C_P(\text{CO}) = 6.97$, $C_P(\text{O}_2) = 7.05$, and $C_P(\text{CO}_2) = 8.96$. Then

$$\Delta C_P = 8.96 - 6.97 - \frac{7.05}{2} = -1.53 \text{ cal/deg}$$

$$\Delta H_{398}^{\circ} = \Delta H_{298}^{\circ} - 1.53(100)$$
$$= -67{,}640 - 153 = -67{,}790$$

The ΔH of reaction is only slightly negative more at this higher temperature.

EXAMPLE 2–7

The enthalpy of vaporization of SO_2 is 5955 cal/mole at its boiling point, 263°K. The heat capacity C_P of the liquid is 20.7 cal/mole-deg, and that of the vapor is 9.3 cal/mole-deg. Calculate the enthalpy of vaporization at 253°K.

$$SO_2(l) = SO_2(g) \qquad \Delta H_{263}^{\circ} = 5955 \text{ cal}$$

$$\Delta H_{253}^{\circ} = \Delta H_{263}^{\circ} + \int_{263}^{253} (9.3 - 20.7) \, dT$$

$$\Delta H_{253}^{\circ} = 5955 + 114 = 6069 \text{ cal/mole}$$

Because of the large difference in heat capacities which exists between liquids and their vapors, the enthalpy of vaporization is generally rather sensitive to changes in temperature.

2–9 EXPLOSIONS AND FLAMES

There is perhaps no better illustration of the usefulness of the state function concept than the problem of calculating the maximum temperature achieved in flames and explosions. We start with a combustible mixture of gases at room temperature and attempt, by using ΔH_{298}° and the heat capacities of reactants and products, to calculate the maximum temperature reached when the gases react. If we make the drastic assumption that the gases react so rapidly they do not lose energy to their surroundings before the reaction is over, then we can see that the maximum temperature reached will depend on the amount of heat produced by the reaction.

However, we have just learned that ΔH and ΔE depend on temperature, and so we are faced with the problem of calculating a final temperature which depends on ΔH, which, in turn, depends on the temperature. Furthermore, as the system reacts it is characterized not only by one temperature, but by a continuously changing temperature, as shown in Fig. 2–8.

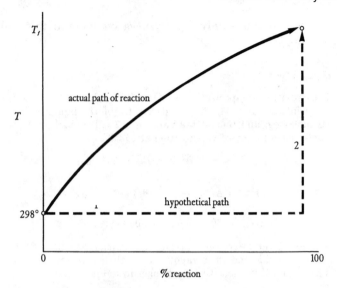

Figure 2–8 Temperature as a function of percentage reaction for an explosion. The dashed curve represents the hypothetical path used to calculate the final temperature.

Although the problem looks hopeless, its solution is incredibly simple if we substitute the two-step process shown in Fig. 2–8 for the actual reaction path. If the reaction is an explosion in a closed container, we first allow the reaction to proceed at constant volume and a constant temperature of 298°K. The heat evolved is stored in a reservoir. Once the reaction is complete, we use this heat to raise the temperature of the products, still at constant volume. When all the heat originally evolved has been returned to the system, the final temperature of the explosion has been reached, since the change in a state function, in this case temperature, is independent of the path taken.

As a numerical example, consider a mixture of 1 mole of hydrogen, $\frac{1}{2}$ mole of oxygen, and 2 moles of nitrogen in a closed bomb of constant volume. The heat involved in the first step is ΔE_{298}. Therefore,

$$H_2(g) + \tfrac{1}{2}O_2(g) = H_2O(g) \qquad \Delta H^\circ_{298} = -57.8 \text{ kcal}$$
$$\Delta H = \Delta E + \Delta n \, RT$$
$$-57{,}800 = \Delta E - \tfrac{1}{2}(1.98)(298)$$
$$\Delta E_{298} = -57{,}500 \text{ cal}$$

In the second step we use this 57,500 cal to heat 1 mole of water vapor and 2 moles of nitrogen at constant volume to a final temperature T. For $H_2O(g)$ and N_2, $C_V = 5.70$ and 4.98, respectively. Therefore,

$$57{,}500 = \int_{298}^{T} (5.70 + 2 \times 4.98) \, dT$$
$$57{,}500 = (15.66)(T - 298)$$
$$T = 3300°K$$

Thus by skillful choice of path and use of the state-function concept, a seemingly insoluble problem is handled with ease. We cannot afford to be too enthusiastic about our answer, however. The assumption that C_V is constant over such an enormous temperature range is surely not correct. An even more serious criticism is that at temperatures above 1000°K, the reaction will not be complete, but considerable amounts of hydrogen and oxygen atoms and OH radicals will be left when the reaction comes to equilibrium. Thus our estimates of the heat produced and the temperature rise are too large. These are not difficulties of principle, however, and they can be remedied by refinement of the calculation.

The calculation of the temperature of a flame is not seriously different from the explosion temperature problem. In a flame, however, the combustion reaction occurs at a constant pressure, so ΔH and C_P are used exactly as were ΔE and C_V in the explosion problem.

2–10   IDEAL-GAS CALCULATIONS

Although our primary object is to learn how to apply thermodynamics to chemical problems, consideration of some simple changes of state for an ideal gas will solidify our understanding of the thermodynamic principles.

Expansion at Constant Pressure

To begin, we have 1 mole of an ideal gas confined in a cylinder with a piston, and we expand this gas from the initial state $(P_1V_1T_1)$ to the final state $(P_1V_2T_2)$. That is, the volume and temperature increase, but the initial and final pressures are the same. What is the work done by the gas in this process? We are totally unable to answer this question, since

$$w = \int_{V_1}^{V_2} P_{ex}\, dV$$

and, although we have specified the initial and final states, nothing has been said about how P_{ex} depends on V. In order to calculate the work done, the *path* followed during the process, as well as the initial and final states, must be known. One way to effect this expansion is to keep the external pressure constant and infinitesimally less than P_1, the gas pressure. That is,

$$P_{ex} = P_1 - dP$$

The behavior of the gas during the expansion is shown as step a in Fig. 2.9. The work done by the system is then

$$w = \int_{V_1}^{V_2} P_{ex}\, dV = \int_{V_1}^{V_2} (P_1 - dP)\, dV$$

Since the product of infinitesimals may be ignored, we can write

$$w = \int_{V_1}^{V_2} P_1\, dV = P_1(V_2 - V_1)$$

Or, by the ideal-gas law

$$w = P_1\left(\frac{RT_2}{P_1} - \frac{RT_1}{P_1}\right) = R(T_2 - T_1)$$

Since $V_2 > V_1$, or $T_2 > T_1$, the work w is a positive quantity, which means that in this process the system has done work on its surroundings.

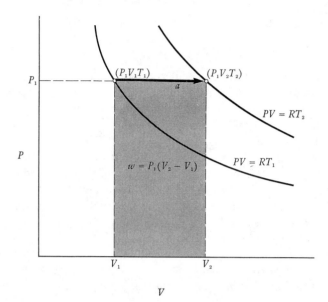

Figure 2-9 The expansion of an ideal gas at constant pressure.

Because the system is at constant pressure throughout the expansion, we can calculate the heat absorbed by the expression

$$q = q_P = \int_{T_1}^{T_2} C_P \, dT = C_P(T_2 - T_1)$$

We have assumed that C_P is a constant independent of temperature. In order to calculate the heat, we had to know *how this process was done.* The knowledge that the process proceeded at constant pressure allowed us to choose C_P as the appropriate heat capacity. Since $T_2 > T_1$, q is positive, which is consistent with our original convention of associating a positive sign with heat *added to* the system.

For ΔH and ΔE we find

$$\Delta H = q_P = C_P(T_2 - T_1)$$
$$\Delta E = q - w = (C_P - R)(T_2 - T_1)$$

But we have already shown that for an ideal gas $C_P = C_V + R$, so

$$\Delta E = C_V(T_2 - T_1)$$

It is particularly important to analyze the manner in which we carried out the expansion. Since $P_{ex} = P_{gas} - dP$, only an infinitesimal net force acted on the piston, and hence the movement of the piston was infinitesimally slow. Such a slow change of state is called a **quasistatic process**. The important characteristic of a quasistatic process is that all state functions change so slowly that the system is at all times in a well-defined equilibrium state. Furthermore, at any time during a quasistatic change, the process can be *stopped and reversed just by making an infinitesimal change in the external conditions.* Quasistatic processes are commonly called **reversible processes** for this reason. In the expansion we have just examined, $P_{ex} = P_1 - dP$. However, if at any time we had set $P_{ex} = P_1 + dP$, the expansion would have ceased and a quasistatic compression would have ensued.

Pressure Change at Constant Volume

Now let us consider a change from the state $(P_1 V_2 T_2)$ to the state $(P_2 V_2 T_1)$. This process is shown as step b in Fig. 2–10. We shall make the change by keeping the volume of the system constant at V_2 and decreasing the temperature and pressure by removing heat from the gas. The process is again carried out quasistatically, so that $P_{ex} = P_{gas}$ and $T_{ex} = T_{gas}$.

We calculate the work from the expression

$$w = \int_{V_2}^{V_2} P_{ex}\, dV = \int_{V_2}^{V_2} P_{gas}\, dV = 0$$

Since there is no displacement, no work can be done by or on the gas. This is just a restatement of the rule of elementary mechanics that a static force can do no work.

Since the process occurs at constant volume,

$$q = q_V = \int_{T_2}^{T_1} C_V\, dT = C_V(T_1 - T_2) = -C_V(T_2 - T_1)$$

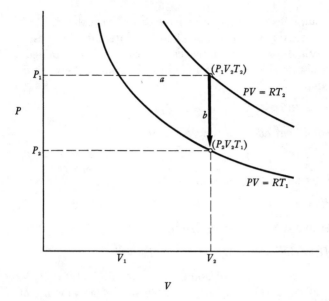

Figure 2-10 The cooling of an ideal gas at constant volume.

We have assumed that C_V is independent of temperature. Since $T_2 > T_1$, the heat has a *negative value*, which indicates that heat is removed from the system, as we anticipated.

The expression for the ΔE of step b is particularly simple.

$$\Delta E = q - w$$
$$\Delta E = C_V(T_1 - T_2) - 0 = -C_V(T_2 - T_1) = q_V$$

Thus the result of step b is that the internal energy of the gas decreases by an amount equal to the heat which the gas gives up to its surroundings.

Calculation of ΔH presents no difficulty if we remember that

$$\Delta H = \Delta E + \Delta(PV) = \Delta E + \Delta(RT)$$
$$= \Delta E + R(T_1 - T_2)$$
$$= -C_V(T_2 - T_1) - R(T_2 - T_1)$$
$$\Delta H = -C_P(T_2 - T_1)$$

This completes the calculations associated with step b.

The total heat, work, and change of internal energy associated with both steps a and b in succession can also be obtained. The total work *done by* the system is just the sum of the work done in each step:

$$w = R(T_2 - T_1) + 0 = R(T_2 - T_1)$$

The total heat *added to* the system is

$$q = (C_V + R)(T_2 - T_1) + C_V(T_1 - T_2) = R(T_2 - T_1)$$

The internal energy change is

$$\Delta E = q - w = R(T_2 - T_1) - R(T_2 - T_1) = 0$$

The enthalpy change is also zero:

$$\Delta H = C_P(T_2 - T_1) - C_P(T_2 - T_1) = 0$$

We find the net result of going from our initial state $(P_1 V_1 T_1)$ to the final state $(P_2 V_2 T_1)$ is the conversion of an amount of heat $R(T_2 - T_1)$ into an equal amount of work, while E and H are unchanged. This calculation illustrates the general conclusion that the *internal energy of an ideal gas depends only on its temperature* and is constant as long as temperature is constant.

Isothermal Expansion

The path that we have used to go from $(P_1 V_1 T_1)$ to $(P_2 V_2 T_1)$ is only one of an infinite number of paths which connect these two states. One obvious alternative is to proceed isothermally, forcing the system to remain at a constant temperature. This process, denoted in Fig. 2–11 by step c, takes the system along the line $PV = RT_1 = $ constant.

We can find ΔE and ΔH for this process quite easily. From the results of our calculations on steps a and b we have learned that there is no difference in internal energy or enthalpy between the initial and final states $(P_1 V_1 T_1)$ and $(P_2 V_2 T_1)$. Since ΔH and ΔE are independent of the path taken between states, we have for process c:

$$\Delta H = 0 \qquad \Delta E = 0$$

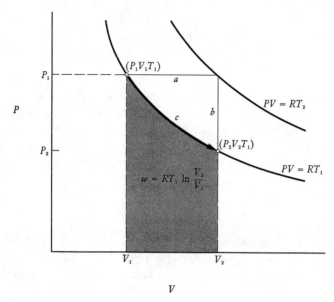

Figure 2-11   The isothermal expansion of an ideal gas.

Now we turn to the calculation of the work done in the isothermal expansion. Again we choose to carry out the expansion reversibly, so that the external pressure and the gas pressure are equal.

$$ w = \int_{V_1}^{V_2} P_{ex}\, dV = \int_{V_1}^{V_2} P_{gas}\, dV $$

Now, in order to perform the integration, we must know how the pressure depends on the volume during the expansion. We are dealing with an ideal gas which, if the process is carried out quasistatically, is at all times in an equilibrium state. Therefore, we can write

$$ P_{gas} = \frac{RT_1}{V} $$

and consequently

$$w = \int_{V_1}^{V_2} P_{\text{gas}}\, dV = \int_{V_1}^{V_2} RT_1 \frac{dV}{V} = RT_1 \ln V \Big|_{V_1}^{V_2}$$

$$w = RT_1 \ln \frac{V_2}{V_1}$$

By the graphical interpretation of integration, the work done is just the area under the curve $P = RT_1/V$ between the volumes V_1 and V_2. Since $V_2 > V_1$, w is a positive quantity, which means that work is done by the system in process c. By combining the calculations of ΔE and w, we can obtain q for process c:

$$\Delta E = q - w = 0$$

$$q = w = RT_1 \ln \frac{V_2}{V_1}$$

The heat q is a positive number, which means heat was added to the gas during the expansion. The result of process c is the conversion of a certain amount of heat into an equivalent amount of work.

Comparison of Expansion Processes

Now, it is profitable to compare the work done by the gas and the heat added to the gas for the two processes which we have used to go from our initial state $(P_1 V_1 T_1)$ to the final state $(P_2 V_2 T_1)$. The work done in steps a and b was $P_1(V_2 - V_1)$, which is greater than the work done along path c, $RT_1 \ln (V_2/V_1)$. A comparison of the areas under the two curves which represent the processes in Fig. 2–11 shows this immediately. Similarly, the amount of heat added to the gas in the combined steps a and b was different from that added in c, even though the initial and final states of the gas were the same. However, the energy and enthalpy changes were the same for both paths. This example again demonstrates that the amount of work and heat involved in a change of state depends on the path followed during the process, but the change in internal energy, enthalpy, or any other state function depends only on the initial and final states of the system. There is a fundamental difference between energy or enthalpy and the quantities heat and work. We can assign an energy or enthalpy to each state of a system, but we cannot say that a system has or contains a certain amount of work or heat. This is simply because both work and heat are defined as processes by which energy can be transferred; once a process is over, the terms "work" and "heat" are not applied to the system. They just are not properties of a system at rest.

Irreversible Expansion

Each of the expansion processes we have studied has been carried out reversibly or quasistatically. Now let us see what differences arise when we expand the same gas irreversibly from $(P_1V_1T_1)$ to $(P_2V_2T_1)$. If we recall that the change in any state function is independent of the path taken between states, we can conclude immediately that ΔH and ΔE for this irreversible expansion are the same as for the corresponding reversible expansion, step c. Therefore, $\Delta H = 0$ and $\Delta E = 0$ for the irreversible isothermal expansion of an ideal gas.

The calculation of the work and heat associated with the irreversible expansion is another matter. Work and heat are not state functions; they do depend on the path taken between states; and they cannot be calculated until we know how the process is done. Suppose the irreversible expansion is accomplished by suddenly dropping P_{ex} from its initial value P_1 to a value equal to P_2. This decrease in P_{ex} occurs without a noticeable volume change. When P_{ex} reaches P_2, the expansion proceeds with P_{ex} held constant at P_2. This path is illustrated in Fig. 2–12. Since P_{ex} is much less than the gas pressure during the expansion, the process is irreversible. The work done by the gas is

$$w = \int_{V_1}^{V_2} P_{ex}\, dV = P_2 \int_{V_1}^{V_2} dV = P_2(V_2 - V_1)$$

From the area under the irreversible path in Fig. 2–12 it is clear that the work done by the gas in the irreversible process is less than in the corresponding reversible expansion which occurs between the same two states. Since $\Delta E = 0$, the heat associated with the irreversible expansion is

$$q = w = P_2(V_2 - V_1)$$

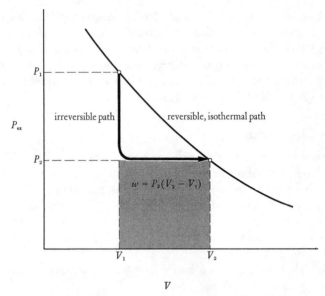

Figure 2–12   The irreversible expansion of an ideal gas.

As we might have expected, q for the irreversible process is different from the heat absorbed by the gas during the reversible expansion.

The examples just completed show once again that for the purpose of calculating the change in a state function we are free to choose the path followed between two states to suit our convenience. However, the work and heat associated with the process will depend on our choice of paths. These facts have been continually emphasized, since they form the basis for the understanding and use of the second law of thermodynamics which we shall treat in the next chapter.

PROBLEMS

1. A steel spring is compressed a distance of 5 cm by a weight of 500 kg. (a) If the acceleration of gravity is 981 cm/sec², what is the internal energy change ΔE of the spring if no heat is lost to the surroundings? (b) If the specific heat of steel is 0.011 cal/g and the spring weighs 5 kg, what temperature rise is produced by the compression?

2. A gas at 1 atm pressure is confined in a cylinder by a movable piston. Suddenly, a constant external force corresponding to 2 atm is applied to the piston, and the gas is compressed from its initial volume of 4 liters to a final volume of 2 liters. (a) What is the work done by the gas? (b) What is the internal energy change if no heat is lost to the surroundings?

3. When 1 mole of water at its boiling point evaporates at a constant pressure of 1 atm, it absorbs 9.7 kcal of heat. Owing to the volume change in going from liquid to vapor, work is done by the system on the surrounding atmosphere. (a) If the volume of a mole of liquid water is 0.019 liter at 373°K and water vapor is an ideal gas, calculate the work done by the evaporation of 1 mole of liquid. (b) What are the values of ΔE and ΔH for this process?

4. At high gas densities the molecules themselves take up an appreciable fraction of the available volume, and an equation of state of 1 mole of gas can be written in the form $P(V - b) = RT$, where b is a constant which is related to the size of the molecules. (a) Calculate the work done when 1 mole of this nonideal gas at 298°K is compressed reversibly and isothermally from 22.4 liters to 0.224 liter if $b = 0.040$ liter. (b) Compare the answer to (a) with the work done in a similar compression of an ideal gas, and explain why the two answers differ as they do.

5. At low temperatures and moderate densities, the attractive forces between molecules become important, and the equation of state of 1 mole of a real gas assumes the form $PV = RT + a/V$, where a is a constant which depends on the strength of the intermolecular attraction. (a) Calculate the work done by 1 mole of this nonideal gas in an isothermal reversible expansion from 0.224 liter to 22.4 liters at 200°K. Assume $a = 3.6$ liter^2-atm. (b) Compare the answer to (a) with the work done in the corresponding expansion of an ideal gas. Why do the two answers differ as they do?

6. Calculate ΔH for the reaction

$$C(\text{graph}) = C(\text{diam})$$

from the following data

$$C(graph) + O_2(g) = CO_2(g) \qquad \Delta H = -94.052 \text{ kcal}$$
$$C(diam) + O_2(g) = CO_2(g) \qquad \Delta H = -94.505 \text{ kcal}$$

7. (a) When 1 mole of methyl alcohol burns at 298°K and constant volume according to the reaction

$$CH_3OH(l) + \tfrac{3}{2}O_2(g) = CO_2(g) + 2H_2O(l)$$

173.65 kcal of heat is evolved. What is ΔH for this reaction? (b) At 298°K, the ΔH_f° of $H_2O(l)$ is -68.32 kcal, and for $CO_2(g)$, ΔH_f° is -94.05 kcal. Calculate the standard enthalpy of formation of liquid methyl alcohol. (c) The enthalpy of vaporization of methyl alcohol is 8.43 kcal/mole. Calculate ΔH_f° for $CH_3OH(g)$.

8. The following reactions are thought to be important in the chemistry of our atmosphere. Using Table 2–2, calculate ΔH° for each of them.

$$N + O_2 = NO + O \qquad NO_2 + O = NO + O_2 \qquad N + N = N_2$$
$$O + O_3 = 2O_2 \qquad\qquad O + O = O_2$$

9. Calculate ΔH° for each of the following reactions

$$Zn(s) + 2H^+(aq) = Zn^{2+}(aq) + H_2(g)$$
$$Ag(s) + H^+(aq) = Ag^+(aq) + \tfrac{1}{2}H_2(g)$$
$$Cu^{2+}(aq) + Zn(s) = Zn^{2+}(aq) + Cu(s)$$

On the basis of these results alone, is it at all surprising that zinc dissolves in acid, while silver does not?

10. When 1 mole of aqueous HCN is neutralized according to the reaction

$$HCN(aq) + OH^-(aq) = H_2O + CN^-(aq)$$

2.9 kcal is evolved at constant pressure and 298°K. Calculate the ΔH of the dissociation of HCN:

$$HCN(aq) = H^+(aq) + CN^-(aq)$$

11. Estimate the ΔH of the following gaseous reactions by using the bond energy table.

$$HCl + \quad \underset{H}{\overset{H}{\diagdown}} C{=}C \underset{H}{\overset{H}{\diagup}} \quad = H{-}\underset{\underset{H}{|}}{\overset{\overset{H}{|}}{C}}{-}\underset{\underset{H}{|}}{\overset{\overset{H}{|}}{C}}{-}Cl$$

$$H{-}\underset{\underset{H}{|}}{\overset{\overset{H}{|}}{C}}{-}\underset{\underset{H}{|}}{\overset{\overset{H}{|}}{C}}{-}OH = H_2O + \quad \underset{H}{\overset{H}{\diagdown}} C{=}C \underset{H}{\overset{H}{\diagup}}$$

$$H{-}\underset{\underset{H}{|}}{\overset{\overset{H}{|}}{C}}{-}H + Cl_2 = HCl + H{-}\underset{\underset{H}{|}}{\overset{\overset{H}{|}}{C}}{-}Cl$$

12. Calculate the enthalpy change when 1 mole of water freezes at constant pressure and a temperature of $-10°C$. At $0°C$, ΔH is -1435 cal/mole, and C_P is 18.0 and 8.8 cal/mole-deg for water and ice, respectively.

13. Find the maximum temperature that might be attained in an acetylene torch flame which burns at constant pressure according to the reaction

$$C_2H_2(g) + \tfrac{3}{2}O_2(g) = 2CO(g) + H_2O(g)$$

Assume that the reaction goes to completion, that the reagents are initially at 300°K, and that there are no energy losses from the flame. (The actual temperature reached in these flames is only 3800°K.)

14. We have said that work is energy transferred by means of a mechanical connection and that heat is energy transferred by means of a temperature difference. Use what you know of kinetic molecular theory to decide what differences there are between the behavior of molecules in a cold gas and the behavior of those in a hot gas. Would it be correct to say that when a hot body and a cold body are brought into contact, energy is really transferred by the "hot" molecules doing work on the "cold" molecules?

III

The Second Law
of Thermodynamics

THE FIRST LAW of thermodynamics expresses the repeated experimental observation that, while it may be transferred between a system and its surroundings, energy is never created or destroyed. Therefore, we propose that the conservation of energy is a minimum requirement which every real process must satisfy. However, a little reflection shows that naturally occurring processes have a feature which is completely inexplicable on the basis of the first law of thermodynamics. This is best understood by consideration of some simple examples.

The first example involves the apparatus shown in Fig. 3–1. Two bulbs are connected by a stopcock; bulb A contains an ideal gas, while bulb B is evacuated. When the stopcock is opened, gas inevitably flows from A to B. It is observed that the system remains at a constant temperature; thus, the process is an isothermal irreversible expansion of an ideal gas. In Chap. 2 we learned that for any isothermal expansion of an ideal gas $\Delta E = 0$. Moreover, there is no mechanical link between our system and its surroundings, so we have $w = 0$, and consequently $q = 0$. Despite the fact that the system is not prodded by an interaction with its

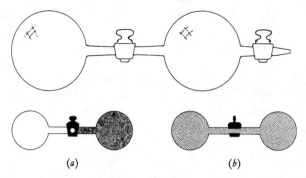

Figure 3-1 Apparatus for the irreversible expansion of a gas. Bulb A contains the ideal gas; bulb B is initially evacuated.

surroundings, the expansion occurs spontaneously, once the stopcock is opened. Now, the reverse process, in which all the molecules in bulb B spontaneously return to bulb A, would also have $q = w = \Delta E = 0$, and according to the first law of thermodynamics this would be a perfectly possible occurrence. However, such a spontaneous concentration of the gas in a totally isolated system has never been observed, and consequently we presume it is impossible.

There are other instances in which systems change in a predictable way but, left to themselves, never return to their original condition. When two blocks of material, one hot and the other cold, are brought together, the temperature of each block changes until the two reach some uniform intermediate temperature. The flow of heat from the hot to the cold block is spontaneous; it proceeds unaided once the blocks are in contact. On the other hand, two isolated bodies in contact and initially at the same temperature have never been observed to depart from temperature uniformity. In order for them to do so, it would be necessary eventually for heat to flow unaided from a cold body to a hot body. This has never been observed, and we state with some confidence that it can never happen. Yet such a process would certainly not violate the first law of thermodynamics, since the energy lost by the cold body would be exactly equal to that gained by the hot body.

As a third example of processes that proceed spontaneously in one direction but not in the reverse, we shall consider a chemical reaction. If hydrogen and oxygen are mixed in a closed container, they react very slowly to form water. This process proceeds unaided, so despite its very small rate it is said to be spontaneous. The introduction of an infinitesimal amount of energy in the form of an electric spark, or the addition of a small amount of finely powdered platinum catalyst, causes the reaction to proceed at a much greater rate. So far as thermodynamics is concerned, however, both the slow reaction and the catalyzed reaction are equally spontaneous processes. As the reaction proceeds, heat is evolved by the system and absorbed by the surroundings. The reverse process, in which the system spontaneously absorbs energy from the surroundings and the water is completely reconverted to hydrogen and oxygen, has never been observed, even though it is allowed by the first law of thermodynamics.

These and many other, similar examples make it clear that natural or spontaneous physical and chemical changes have a *direction* which cannot be explained by the first law of thermodynamics. Repeated observations of simple spontaneous processes show that their universal property is that systems tend to move from a nonequilibrium state toward a state which satisfies the equilibrium criteria; once in an equilibrium state, systems do not change unless disturbed by some external agent. In our first example, the gas is in a nonequilibrium state immediately after the stopcock connecting the two bulbs is opened, since the pressure is not uniform throughout the system. Similarly, the system formed immediately after two bodies at different temperatures are brought together is in a nonequilibrium state, since its temperature is not uniform.

The processes by which the systems approach equilibrium are irreversible processes, since they cannot be reversed by any infinitesimal changes in external conditions. The irreversible flow of gas eventually results in pressure uniformity, and the irreversible flow of heat leads to temperature uniformity. Once in these equilibrium states, the systems do not change, as long as they are undisturbed. Thus it appears that prediction of the occurrence and direction of some changes is a simple matter: all we need do is ascertain whether the first two criteria for an equilibrium state are satisfied; if not, the changes that will inevitably ensue are those which will bring about equilibrium. Heat will flow if temperature is not uniform, and matter will flow if pressure is not uniform.

But what of chemical processes? Just as we are able to use the nonuniform temperature of a bar of metal to predict the change that will occur, so we want to be able to use the properties of a mixture of hydrogen and oxygen to predict that these gases will react to form water. Just as we know that uniform temperature and pressure are necessary for equilibrium in simple physical systems, so we want to know what concentration conditions correspond to equilibrium in a chemical system. What we seek is a property of matter, a state function, which can be used to predict the direction of spontaneous change and the eventual equilibrium state of all systems.

The second law of thermodynamics identifies a new state function and provides a criterion for identifying the equilibrium state of a system. There are several equivalent ways of stating the second law, and one of the most useful is the following:

There is a quantity S, called **entropy**, which is a function of the state of a system. In an irreversible process the entropy of the universe increases. In a reversible process the entropy of the universe remains constant. At no time does the entropy of the universe decrease.

These statements summarize an enormous accumulation of experimental information concerning the properties of entropy. But how is entropy defined and measured? The general mathematical expression for the difference in entropy ΔS between states 1 and 2 of a system is

$$\Delta S = \int_1^2 \frac{dq_{rev}}{T} \qquad (3\text{--}1)$$

This equation says that, to compute ΔS, move the system from state 1 to state 2 *by means of a reversible path*. For each infinitesimal step of the reversible path, compute dq/T; and the sum of these quantities is ΔS. The expression is very simple for a process done at constant temperature:

$$\Delta S = \int_1^2 \frac{dq_{\text{rev}}}{T} = \frac{1}{T} \int_1^2 dq_{\text{rev}} = \frac{q_{\text{rev}}}{T} \quad \text{(constant } T\text{)}$$

It is very important to realize that the entropy change must *always* be computed by taking the *system* from state 1 to state 2 by a *reversible path*. However, since S is a function of state, the ΔS of the system is *independent of the path taken* and depends only on the initial and final states of the system. Although these two statements may seem contradictory, they are not, since $dq_{\text{rev}}/T \neq dq_{\text{irrev}}/T$. The situation here is similar to that encountered in the computation of ΔH; ΔH is independent of the path, but it is equal to q only when the process is carried out at constant pressure. The entropy change is also independent of the path, but it is equal to $\int dq/T$ only *when the process is carried out reversibly*. It is the quantity $\int dq/T$ which depends on how the process is done, and not the ΔS of the system.

The second law of thermodynamics is a general principle that summarizes the results of a great many experiments. Owing to this generality, it may not be clear how the second law provides a criterion for the spontaneity of chemical and physical processes. Therefore, before taking up the details of calculating entropy changes, we offer an outline of how the second law of thermodynamics is to be applied.

First we recognize that a spontaneous change is an irreversible process, and the second law states that in all irreversible processes the entropy of the system and its surroundings (the universe) increases. Then, if we are confronted with a system in a given state and we are asked whether a certain change in its state will be spontaneous, we have only to compute the entropy change which will accompany the contemplated change. If ΔS for the system and surroundings proves to be positive, the contemplated process is spontaneous. If ΔS is zero or negative for any and all conceivable changes, then the system will be unable to change spontaneously and is therefore at equilibrium. Clearly, ability to calculate entropy changes is required before we can put this plan into effect, and it is to this problem that we now turn.

3–1   ENTROPY CALCULATIONS

In Chap. 2, we emphasized that the change in a state function depends only on the initial and final states of the system and not on the path followed between states. We shall put this fact to use in the calculation of entropy changes, for it will always be necessary to replace any actual *irreversible* path by an equivalent, if hypothetical, *reversible* path, in order to effect the calculation.

Reversible Isothermal Expansion

Let us calculate the entropy change of n moles of an ideal gas when it is expanded isothermally from V_1 to V_2. Regardless of how this expansion is to be carried out in practice, according to Eq. (3–1), ΔS must be computed by carrying the system from its initial state to its final state by means of a *reversible path*. For a reversible, isothermal expansion we have already learned that

$$q = w = \int P_{\text{ex}}\, dV = \int P_{\text{gas}}\, dV = nRT \ln \frac{V_2}{V_1}$$

Therefore, the entropy change of the gas is

$$\Delta S = \frac{q_{\text{rev}}}{T} = \frac{nRT}{T} \ln \frac{V_2}{V_1} = nR \ln \frac{V_2}{V_1} \qquad (3\text{–}2)$$

Notice carefully that the entropy of the system increases as the volume increases. This correlation between the volume and entropy applies to all types of systems, but the particular algebraic relation shown in Eq. (3–2) is appropriate only for ideal gases.

The second law of thermodynamics states that in a reversible process the entropy of the universe remains constant. In any problem the only part of the universe which need concern us is the system and its immediate surroundings, since only these parts will change. The surroundings in the isothermal expansion of an ideal gas are a thermostatic bath which maintains a constant temperature and delivers heat to the system as the expansion proceeds. When the system *absorbs* q_{rev} units of heat in the reversible expansion, the surroundings *lose* an equal amount. Therefore, the entropy change of the surroundings is

$$\Delta S_{surr} = \frac{-q_{rev}}{T}$$

Here we are thinking of q_{rev} as a positive number, since it is the heat absorbed by the system during the expansion. Therefore, the entropy of the surroundings decreases as it delivers heat to the system. Does this decrease in the entropy of the surroundings constitute a violation of the second law of thermodynamics? Not at all. The entropy of any part of the universe may decrease, as long as there is a simultaneous, compensating increase in the entropy of some other part.

The total entropy change is the sum of the individual changes in the system and surroundings:

$$\Delta S_{total} = \Delta S_{syst} + \Delta S_{surr}$$

$$\Delta S_{total} = \frac{q_{rev}}{T} - \frac{q_{rev}}{T} = 0 \quad \text{(reversible process)}$$

The total entropy change accompanying this reversible process is zero, as the second law of thermodynamics requires.

Irreversible Isothermal Expansion

According to the second law, the entropy of a system is a function only of its state. Therefore, the difference in the entropy of two given states of a system is always the same, regardless of how the system is taken from one state to the other. This fact allows us to calculate the entropy change of an ideal gas which expands *irreversibly from* V_1 to V_2 by using the results of the corresponding *reversible* expansion. Since the entropy change of the system depends only on its initial and final states, we can write immediately

$$\Delta S_{syst} = nR \ln \frac{V_2}{V_1}$$

just as in the reversible case.

How, then, does the irreversible expansion differ from the reversible process? To find out, we must consider the entropy change of the surrounding thermostat. The only role of the thermostat is to deliver the correct amount of heat needed to keep the system at a constant temperature. Suppose the irreversible expansion is accomplished by opening a stopcock which allows the gas to expand from V_1 into an evacuated bulb so that its final volume is V_2. Then, as we showed at the beginning of this chapter,

$$\Delta E = 0 \qquad q_{irrev} = w_{irrev} = 0$$

since in this case the system does no work on its surroundings. The system absorbs no heat; therefore, the thermostat loses none, and for the entropy change of the surroundings we have

$$\Delta S_{surr} = \frac{q}{T} = \frac{0}{T} = 0$$

It may seem that in calculating the entropy change of the thermostat we have violated the rule of always using q_{rev} in the computation of ΔS, but in fact we have not. The thermostat does not "know" whether the expansion is reversible or not, and we can say it *always* delivers the required heat to the system in a reversible manner, as long as the system and the thermostat are at the same temperature. Therefore, to compute the entropy change of the thermostat, we need only divide the heat it absorbs by its temperature.

The total entropy change that is a result of the irreversible process is

$$\Delta S_{total} = \Delta S_{syst} + \Delta S_{surr}$$
$$\Delta S_{total} = nR \ln \frac{V_2}{V_1} + 0$$

Since $V_2 > V_1$, the entropy of the universe as a result of the irreversible expansion is greater than zero, as is consistent with the second law of thermodynamics. These examples show how essential it is to consider the total entropy change of the system and its surroundings before attempting to conclude whether or not a process is spontaneous.

The General Isothermal Process

Although we have demonstrated that ΔS of the system and its surroundings is positive for a *particular* irreversible isothermal expansion of an ideal gas, it is useful to proceed further and show that this is true for *any* irreversible isothermal expansion of any substance. Our argument involves a comparison between the work done by the system in the reversible and irreversible expansions. If we write P_{int} for the equilibrium pressure of the system, in the case of a reversible expansion we have

$$P_{ex} = P_{int}$$
$$w_{rev} = \int P_{ex}\, dV = \int P_{int}\, dV$$

But when the expansion is irreversible, we can only write

$$P_{ex} < P_{int}$$
$$w_{irrev} = \int P_{ex}\, dV < \int P_{int}\, dV = w_{rev}$$

Therefore

$$w_{irrev} < w_{rev} \qquad\qquad (3\text{-}3)$$

That is, since P_{ex} in the irreversible case is always lower than P_{ex} (or P_{int}) in the reversible case, the work done by the system in the irreversible expansion must always be less than that done in the reversible case. The application of these ideas to the special case of an isothermal expansion of an ideal gas is shown in Fig. 3–2. However, Eq. (3–3) is quite general for all isothermal processes and is not just restricted to ideal gases.

Now we must compare the heat for the two types of processes. From the first law we have

$$q_{rev} = \Delta E + w_{rev}$$
$$q_{irrev} = \Delta E + w_{irrev}$$

Subtracting, we find

$$q_{rev} - q_{irrev} = w_{rev} - w_{irrev}$$

But since

$$w_{irrev} < w_{rev}$$

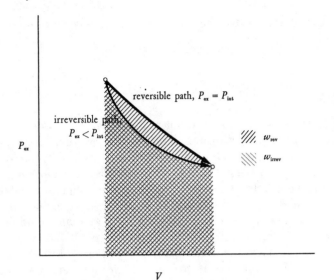

Figure 3–2 **Comparison of the work done in reversible and irreversible expansions between the same states. The area under the reversible path is always greater than under the irreversible path.**

we can write

$$q_{rev} - q_{irrev} = w_{rev} - w_{irrev} > 0$$

$$q_{rev} > q_{irrev}$$

The heat absorbed by the system in an isothermal reversible process is always greater than in any corresponding irreversible process. Since the entropy change in the system is always q_{rev}/T and the entropy change of the surroundings is numerically equal to $-q_{irrev}/T$, we find for irreversible isothermal processes

$$\Delta S_{total} = \frac{q_{rev}}{T} - \frac{q_{irrev}}{T} > 0 \qquad \text{(irreversible)}$$

while for reversible processes the entropy change of the surroundings is $-q_{rev}/T$, so

$$\Delta S_{total} = \frac{q_{rev}}{T} - \frac{q_{rev}}{T} = 0 \qquad \text{(reversible)}$$

This completes our demonstration of the second law for isothermal processes.

Temperature Dependence
of Entropy

Now let us turn to the problem of the calculation of entropy differences between states of different temperature. Again, we must imagine ourselves taking a system from its initial state to its final state by a reversible path. A reversible change in temperature can be effected by slowly changing the temperature of the surroundings, so that the temperature of the surroundings differs from the temperature of the system by no more than an infinitesimal amount dT. This infinitesimal temperature difference causes an infinitesimal amount of heat dq_{rev} to be absorbed by the system at each of the temperatures between the initial and final values. For a change taking place at constant pressure, combination of the definitions of entropy and heat capacity yields

$$dS = \frac{dq_{rev}}{T}$$

$$nC_p = \frac{dq}{dT} \qquad \text{(constant pressure)}$$

$$dq = nC_p \, dT$$

$$dS = \frac{nC_P}{T} \, dT$$

$$\Delta S = \int_{T_1}^{T_2} \frac{nC_P}{T} \, dT \qquad (3\text{–}4)$$

For the special case in which C_P is independent of temperature, integration of Eq. (3–4) gives us

$$\Delta S = nC_P \ln \frac{T_2}{T_1} \qquad \text{(constant pressure)} \quad (3\text{–}5)$$

A similar procedure for constant-volume processes results in the expression

$$\Delta S = nC_V \ln \frac{T_2}{T_1} \qquad \text{(constant volume)} \quad (3\text{–}6)$$

Equations (3–5) and (3–6) pertain to processes in which the system is heated or cooled reversibly from T_1 to T_2; however, since ΔS of the system is independent of the path, they also give the entropy change of a system whose temperature has changed irreversibly. Let us use them to calculate the entropy change which results when a hot body T_h is brought in contact with a cold body T_c. Suppose the bodies are both composed of 1 mole of the same material; then the final temperature T which they reach is found by saying:

heat lost by hot body = heat gained by cold body

$$C_P(T_h - T) = C_P(T - T_c)$$

$$T = \frac{T_h + T_c}{2}$$

Since the hot body cools from T_h to T, its entropy change is, by Eq. (3–5),

$$\Delta S_h = C_P \ln \frac{T}{T_h}$$

For the cold body

$$\Delta S_c = C_P \ln \frac{T}{T_c}$$

Since the two bodies interact only with each other, the total entropy change of the universe is

$$\Delta S_t = \Delta S_h + \Delta S_c = C_P \left(\ln \frac{T}{T_h} + \ln \frac{T}{T_c} \right)$$

$$\Delta S_t = C_P \ln \frac{T^2}{T_h T_c}$$

If $T^2 \geqslant T_h T_c$, then the entropy change of the universe is equal to or greater than zero. We must therefore show that for all finite temperatures

$$T^2 = \left(\frac{T_h + T_c}{2} \right)^2 \geqslant T_h T_c$$

This is true if

$$T_h{}^2 + T_c{}^2 + 2T_hT_c \geqslant 4T_hT_c$$

or

$$T_h{}^2 - 2T_hT_c + T_c{}^2 \geqslant 0$$
$$(T_h - T_c)^2 \geqslant 0$$

Therefore, $\Delta S > 0$ and the process is irreversible as long as $T_h > T_c$. When the two initial temperatures are equal, $\Delta S = 0$ and no irreversible flow of heat results.

3–2 EQUILIBRIUM IN ISOLATED SYSTEMS

The second law of thermodynamics says that the entropy of a *system and its surroundings* must increase in an irreversible process, or otherwise must remain constant. But suppose we have an isolated system that can in no way exchange energy with the rest of the universe. Since such a system has no surroundings, we can only conclude that the entropy of the isolated *system* alone either increases or stays constant. That is, as the isolated system undergoes an irreversible change as its entropy increases, and when the system eventually reaches equilibrium, its entropy stops increasing and remains constant.

Thus we can say that in an isolated system, where energy is constant, the condition for equilibrium is that *the entropy reach some maximum value* beyond which it cannot increase. To illustrate the application of this general criterion for equilibrium in isolated systems, we need only refer to the expansion of an ideal gas into an evacuated bulb. This system can be completely isolated, so that q, w, and ΔE are zero. Once the stopcock is opened, the volume occupied by the gas increases from its initial value V_1 until its volume is V_2, the total volume of the two bulbs.

As we saw earlier, the entropy change of the system is

$$\Delta S = nR \ln \frac{V_2}{V_1}$$

so the entropy increases as the gas volume increases, and this entropy increase continues until the volume of the gas is limited by the walls of the vessel. At this point the entropy has reached a maximum beyond which it cannot go, and the system has reached equilibrium.

The flow of heat from a hot to a cold body also serves to illustrate that equilibrium in an isolated system is reached when entropy can no longer increase. As we saw earlier, when 1 mole of material at T_c and 1 mole of material at T_h are brought together and allowed to reach a uniform intermediate temperature T, the entropy of this composite isolated system increases by an amount

$$\Delta S = C_P \ln \frac{T^2}{T_h T_c}$$

We can regard the entropy of the system as increasing until the temperature of the system becomes uniform. Then, as long as the system is isolated, there is no conceivable change that can cause the entropy to increase further. Thus when the temperature is uniform, the entropy of the system has reached a maximum, and the system itself has reached equilibrium. Pressure and temperature uniformity are two of the requirements for equilibrium which were given in Chap. 1. This brief analysis has indicated that they may be thought of as consequences of the more general principle that entropy must be a maximum in isolated systems at equilibrium. The entropy-maximum principle also provides the basis for finding the conditions for chemical equilibria. Before demonstrating this, we must first find how entropy depends on the nature and composition of chemical systems.

3-3 MOLECULAR INTERPRETATION OF ENTROPY

As stated earlier, the molecular interpretation of macroscopic properties such as pressure, temperature, energy, and entropy lies outside the scope of thermodynamics. However, in this section we shall try to make entropy a more meaningful concept by interpreting entropy changes in terms of molecular behavior.

We mentioned in Chap. 1 that in specifying a thermodynamic state we ignore the values of the coordinates and velocities of each individual atom as being unnecessary for a macroscopic description of a system. We must remember, however, that any observed macroscopic property is a consequence of the motion and position of these atomic particles. Now let us suppose that at any instant we are able to know the position and velocity of each atom in a mole of gas. A list of these positions and velocities would completely describe the gas in its microscopic detail; therefore, such a list is said to specify the *microscopic state* of the system. A different list of velocities and positions of the atoms would correspond to a different microscopic state.

The microscopic state of a system changes with astounding rapidity. In a gas, molecules are changing their positions at a rate of approximately 10^5 cm/sec. In addition, each molecule collides with another approximately every 10^{-9} sec when the gas pressure is 1 atm, and every collision leads to a change in velocity. Therefore, even during the time it takes to measure a macroscopic property of a gas, such as its pressure, the microscopic state of the gas must change an enormous number of times.

Despite this furious molecular activity, the macroscopic properties of a system at equilibrium are constant. Our only conclusion can be that there is a very great number of microscopic states which are consistent with or which can cause any macroscopic state we observe. Entropy is a measure of how many different microscopic states are consistent with a particular macroscopic state. Macroscopic states which are caused by many different microscopic states have large entropies. A careful analysis of this idea shows that the relation between entropy and the number of microscopic states Ω is

$$S = \frac{R}{N} \ln \Omega$$

where R is the gas constant and N is Avogadro's number. At room temperature, the number of microscopic states available to a typical system is in the neighborhood of $10^{10^{23}}$, an extremely large number.

In order to understand these ideas better, let us consider a deck of cards, which provides a useful analogy to the usual thermodynamic system. The deck has two recognizable macroscopic states: the shuffled condition in which it is impossible to predict the value of, for example, the tenth card from the top of the deck and the perfectly ordered state in which the cards are arranged in some known order that allows the prediction of the value of any card from a knowledge of its position in the deck. The actual order of the cards is the microscopic state of the deck.

Of course, there are many sequences, or microscopic states, which correspond to the shuffled macroscopic state, whereas there is only one microscopic state of perfect order. Therefore, we can say the shuffled state has higher entropy than the unshuffled, or ordered, state. Suppose we start with the deck perfectly ordered and then mix the cards. The entropy of the system increases and the deck moves toward a state of equilibrium in which the cards are in one of the many possible random sequences. The fundamental reason why the shuffled state is found after mixing the cards is that there are many microscopic states that are consistent with the shuffled condition and only one that is consistent with the ordered state. This is exactly the reason that the entropy of an isolated physical system increases until it reaches a maximum value at equilibrium. The equilibrium state has associated with it many more microscopic states than have nonequilibrium states. Therefore, in the long run the system will be found in the equilibrium state simply because it is more probable than the nonequilibrium states.

The deck-of-cards analogy suggests another useful interpretation of entropy. A perfectly ordered deck has only one microscopic state and hence low entropy, whereas a shuffled or chaotic deck may be in one of many microscopic states and therefore has high entropy. Thus entropy appears to be a measure of the disorder or chaos of the deck of cards. The situation in molecular systems is similar. The number of microscopic states available to a highly ordered molecular system, such as a crystal, is small, and hence the entropy of such a system is small. In contrast, when molecules are free to move throughout the entire volume of a large vessel, there are a great number of positions and velocities available to them, and the system has a high entropy. Since we are then unable to say with any certainty where one gaseous molecule is with respect to the others, we feel such a situation is characterized by molecular disorder. This interpretation supplies the reason that entropy increases as volume and temperature increase. As volume increases, the number of positions available to a molecule increases, and the chance that several molecules will be found together in an *ordered* cluster decreases. Thus entropy increases with increasing volume.

To understand the effect of temperature on entropy, we have to consider the order among the speeds of the molecules. If all molecules moved at exactly the same rate, we would have a state of perfect order of the speeds. Elementary kinetic theory shows that the speeds of molecules in a real gas, far from being the same, are distributed from zero to very high values, as shown in Fig. 3–3. Since it is not possible to predict with absolute certainty what the speed of a particular molecule will be, we must consider the speeds of the molecules to be disordered. The amount of disorder increases as temperature increases, as shown in Fig. 3–3. That is, at lower temperatures most of the molecules

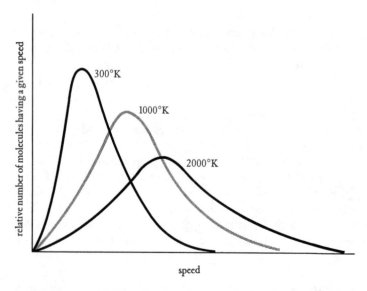

Figure 3–3 Curves showing the spread of molecular speeds in a gas as a function of temperature.

have speeds near the most probable speed, which corresponds to the maximum in the distribution curve in Fig. 3–3. At higher temperatures, the range of the speeds which a molecule might be likely to have is more extended, and more of the molecules have speeds which differ greatly from the most probable value. This means that when the gas is at high temperature, we can predict the speed of a particular molecule with less certainty than when the gas is cold. Correspondingly, the speeds at high temperature can be thought of as more disordered than at low temperature.

The following calculations of the entropy changes which accompany phase changes demonstrate that entropy increases with increasing molecular disorder. When a solid melts, we have

$$\Delta S = \frac{q_{rev}}{T} = \frac{\Delta H_f}{T_f}$$

where ΔH_f is the enthalpy of fusion and T_f is the melting temperature. For the transformation of a solid to a liquid, ΔH is positive, and so the entropy of a substance increases when it melts. In a solid, the molecules must occupy fixed sites in the crystal lattice, while in a liquid the location of one molecule relative to all others is much less restricted. The increased number of allowed molecular positions in the liquid results in increased chaos, disorder, and entropy. When solid SO_2 melts, its entropy change is

$$SO_2(s) \rightarrow SO_2(l)$$

$$\Delta S = \frac{1769 \text{ cal/mole}}{197°K} = 8.95 \text{ cal/mole-deg}$$

$$= 8.95 \text{ eu (entropy units)}$$

The heat of fusion of mercury is 549 cal/mole, and its melting temperature is 234°K. Therefore

$$Hg(s) \rightarrow Hg(l)$$

$$\Delta S = \frac{549}{234} = 2.34 \text{ cal/mole-deg}$$

$$= 2.34 \text{ eu}$$

The entropy change upon melting of most solids lies in this range from 2 to 9 eu, which shows that the increase in molecular disorder or freedom upon melting is roughly comparable for all substances.

For the evaporation of a liquid at its boiling temperature T_b

$$\Delta S = \frac{q_{rev}}{T} = \frac{\Delta H_{vap}}{T_b}$$

where ΔH_{vap} is the enthalpy of vaporization. Since the transformation of liquid to its vapor requires heat, ΔH_{vap} and ΔS_{vap} are always positive. For the evaporation of 1 mole of *n*-hexane at its boiling point

$$C_6H_{14}(l) \rightarrow C_6H_{14}(s)$$

$$\Delta S = \frac{6500}{308} = 21.1 \text{ cal/mole-deg}$$

For the vaporization of ether

$$(C_2H_5)_2O(l) \rightarrow (C_2H_5)_2O(g)$$

$$\Delta S = \frac{6500}{308} = 21.1 \text{ cal/mole-deg}$$

Notice that the increase in entropy upon vaporization is considerably larger than the entropy of fusion. This indicates that the increase in disorder in going from a liquid to its vapor is larger than the change in going from a solid to a liquid. This is entirely consistent with our molecular picture of these three states of matter. A solid approaches a state of perfect molecular order; a liquid is slightly less ordered; and a gas approaches complete molecular chaos.

It is useful to notice that the entropy of vaporization of most liquids is approximately 21 eu. This observation, called Trouton's rule, provides a simple way of estimating ΔH_{vap} for a liquid if its boiling point is known. Since

$$\Delta S_{vap} \simeq 21 \text{ eu}$$
$$\Delta H_{vap} \simeq 21 T_b$$

The molecular significance of Trouton's rule is that since ΔS_{vap} is always approximately 21 eu, the increase in molecular chaos or disorder upon vaporization is almost the same for all liquids.

3-4 EVALUATION OF ABSOLUTE ENTROPIES

We have shown how to calculate the difference in entropy of two states of a system, but we have not as yet assigned an absolute value to the entropy of any state. The molecular interpretation of entropy suggests the means of determining absolute entropies. Since the entropy is related to the number of microscopic states Ω available to the system by the equation

$$S = \frac{R}{N} \ln \Omega$$

we can assign zero entropy to the macroscopic state having $\Omega = 1$. In a perfect crystal at absolute zero there is only one distinguishable microscopic state or mechanical condition. Each atom occupies one of the crystal lattice sites and has the lowest possible energy. This argument, and a wealth of supporting experimental data, form the basis of the **third law of thermodynamics:**

> The entropy of a perfect crystal of each element and compound is zero at the absolute zero of temperature.

As a consequence of the third law, we can evaluate the absolute entropy S of a substance by means of the expressions

$$S - S_0 = \int_0^T \frac{dq_{\text{rev}}}{T} \quad \text{and} \quad S_0 = 0 \qquad (3\text{--}7)$$

If we wish to determine the entropy at room temperature of a solid such as graphite, which does not undergo a phase transition between $T = 0$ and $T = 298°K$, we need only make the substitution

$$dq_{\text{rev}} = C_P \, dT \qquad \qquad \text{(constant } P\text{)}$$

$$S = \int_0^T \frac{C_P \, dT}{T} = \int_0^T C_P \, d(\ln T) \quad \text{(constant } P\text{)} \quad (3\text{--}8)$$

We remarked in an earlier section that C_P (and C_V) is a function of temperature. Thus we can integrate Eq. (3–8) only if we know how C_P depends on temperature. The general form of its temperature dependence is shown in Fig. 3–4. At high temperatures,

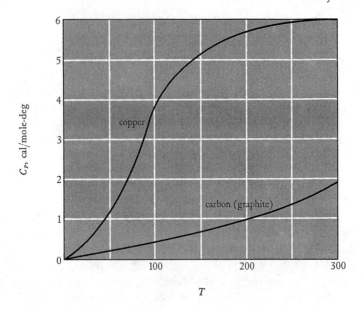

Figure 3–4   Heat capacity as a function of temperature.

C_P of a solid approaches the value $3R$ per mole of atoms, while at low temperatures C_P approaches zero. There is no convenient analytical expression which is a satisfactory representation of C_P over the complete temperature range. However, once C_P has been measured as a function of temperature, Eq. (3–8) can be integrated graphically. One procedure is to plot C_P/T as a function of T and determine the area under the curve. An equivalent procedure is to find the area under a curve of C_P as a function of $\ln T$. These two types of plots are shown in Fig. 3–5.

If a phase transition such as melting occurs at some temperature T_f before the substance reaches 298°K, the evaluation of Eq. (3–7) then leads to three terms:

$$S = \int_0^T \frac{dq_{\text{rev}}}{T} \tag{3-7}$$

$$S = \int_0^{T_f} \frac{C_P}{T}\, dT + \frac{\Delta H_f}{T_f} + \int_{T_f}^T \frac{C_P'\, dT}{T}$$

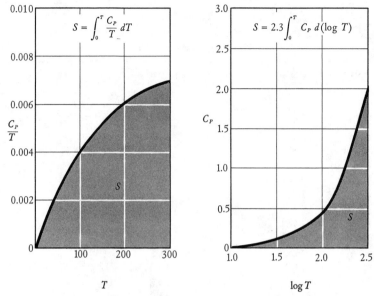

Figure 3-5 (a) C_P/T as a function of T for graphite. The area under the curve is equal to the entropy. (b) C_P as a function of log T for graphite. The area under the curve, multiplied by 2.303, is equal to the entropy.

Here C_P is the heat capacity of the solid, T_f is its melting temperature at 1-atm pressure, and C_P' is the heat capacity of the liquid. The heat capacity integrals are to be evaluated graphically as before, and the entropy of melting is calculated from the measured ΔH_f and T_f. The contribution of any other phase transitions which occur must be included in a similar manner.

Table 3-1 gives some values of absolute entropies of compounds and elements in their standard states. Study the table carefully and notice how closely entropy and molecular structure are related. Remember that entropy is a measure of the chaos or freedom which atoms have in a particular state. Solid substances have an entropy which is generally about 10 to 15 eu *per mole of atoms*. Higher values occur in soft substances in which the atoms are held rather loosely to their crystal lattice positions,

Table 3-1
Absolute Entropies
$S°$, cal/mole-deg, at 298°K

Solid elements		Solid compounds		Liquids	
Ag	10.20	AgCl	23.00	Br_2	36.4
Al	6.77	AgBr	25.6	H_2O	16.73
B	1.7	AgI	27.6	Hg	18.17
Ba	15.1	BaO	16.8		
C(graph)	1.37	$BaCO_3$	26.8		
C(diam)	0.6	$BaSO_4$	31.6		
Ca	9.95	CaO	9.5		
Cu	7.97	$Ca(OH)_2$	17.4		
Fe	6.49	$CaCO_3$	22.2		
I_2	27.76	CuO	10.4		
Na	12.2	Cu_2O	24.1		
S(rh)	7.62	Fe_2O_3	21.5		
Si	4.51	SiO_2	10.0		
Zn	9.95	ZnO	10.5		
		ZnS	13.8		

Monatomic gases		Gaseous diatomic molecules		Gaseous polyatomic molecules	
He	30.13	H_2	31.21	H_2O	45.1
Ne	34.95	D_2	34.6	CO_2	51.1
Ar	36.98	F_2	48.6	SO_2	59.4
Kr	39.19	Cl_2	53.3	H_2S	49.1
Xe	40.53	Br_2	58.6	NO_2	57.5
H	27.39	CO	47.3	N_2O	52.6
F	37.92	NO	50.3	NH_3	46.0
Cl	39.46	N_2	45.7	O_3	56.8
Br	41.80	O_2	49.0		
I	43.18	HF	41.5		
N	36.61	HCl	44.6		
C	37.76	HBr	47.4		
O	38.47	HI	49.3		

Aqueous ions			
H^+	0.0	OH^-	-2.52
Na^+	14.4	Cl^-	13.2
K^+	24.5	Br^-	19.3

Table 3-1 (continued)

Ag^+	17.7	I^-	26.1
Ba^{2+}	3.0	HS^-	14.6
Ca^{2+}	-13.2	S^{2-}	5.3
Cu^{2+}	-23.6	SO_4^{2-}	4.1
Zn^{2+}	-25.4	CO_3^{2-}	-12.7

Organic compounds
Gases

Methane, CH_4	44.5	Ethylene, C_2H_4	52.45
Ethane, C_2H_6	54.8	Acetylene, C_2H_2	49.99
Propane, C_3H_8	64.5	1-Butene, C_4H_8	73.48
n-Butane, C_4H_{10}	74.10	cis-2-Butene, C_4H_8	71.9
Isobutane, C_4H_{10}	70.42	$trans$-2-Butene, C_4H_8	70.9
n-Pentane, C_5H_{12}	83.4	Isobutene, C_4H_8	70.2
Isopentane, C_5H_{12}	82.1	1,3-Butadiene, C_4H_6	66.62
Neopentane, C_5H_{12}	73.2	Methyl chloride, CH_3Cl	55.97

Liquids

Methanol, CH_3OH	30.3	Benzene, C_6H_6	41.3
Ethanol, C_2H_5OH	38.4	Chloroform, $CHCl_3$	48.5
Acetic acid, CH_3COOH	38.2	Carbon tetrachloride, CCl_4	51.2

while hard substances such as diamond, quartz, silicon, and boron, which have rigid crystal lattices, have entropies of 5 eu per mole of atoms, or less. Liquids, with their increased disorder, have higher entropies than solids. Among the gases, notice that all diatomic molecules have nearly the same entropy, with hydrogen-containing molecules a little lower than the others.

This correlation between molecular structure and entropy is not accidental; it is possible to calculate the entropy of a gaseous molecule just by knowing certain of its mechanical properties. However, the exact relation between molecular structure and entropy is part of the subject of statistical mechanics and lies outside the scope of classical thermodynamics.

It is important to realize that both elements and compounds have entropies which are greater than zero at 298°K and that the absolute entropy of a compound *is not* its entropy of formation from the elements. The entropy of formation of a compound is calculated· by

$$\Delta S_f^\circ = S^\circ(\text{compound}) - \Sigma \, S^\circ(\text{elements})$$

The entropy change in any reaction is

$$\Delta S^\circ = \Sigma \, S^\circ(\text{products}) - \Sigma \, S^\circ(\text{reactants})$$

where we refer to absolute entropies on the right-hand side of the equation.

EXAMPLE 3–1

Calculate the standard entropy change which accompanies the formation of 1 mole of gaseous HCl from its elements.

$$\tfrac{1}{2}H_2(g) + \tfrac{1}{2}Cl_2(g) = HCl(g)$$
$$\Delta S^\circ = S^\circ(HCl) - \tfrac{1}{2}S^\circ(H_2) - \tfrac{1}{2}S^\circ(Cl_2)$$
$$\Delta S^\circ = 44.6 - 15.6 - 26.6 = 2.4 \text{ eu}$$

Since the molecular structures of the products and reactants are similar, the entropy change is small.

EXAMPLE 3–2

Calculate the standard entropy change which accompanies the decomposition of calcium carbonate by the reaction

$$CaCO_3(s) = CaO(s) + CO_2(g)$$
$$\Delta S^\circ = S^\circ(CaO) + S^\circ(CO_2) - S^\circ(CaCO_3)$$
$$\Delta S^\circ = 9.5 + 51.1 - 22.2$$
$$\Delta S^\circ = 38.4 \text{ eu}$$

The entropy change is large and positive. The product molecules represent a more disordered state of the system because the CO_2 molecules are not located in fixed positions, as was the case when these atoms were part of the carbonate ion in the $CaCO_3$ crystal.

3–5   FREE ENERGY

The condition of maximum entropy is a convenient criterion for equilibrium in systems which are isolated from their surroundings. However, when we wish to treat systems which are not isolated, we must consider entropy changes not only of the system itself but of its surroundings as well. Most of the systems in which chemists are interested are not isolated; they readily exchange energy with their surroundings. It is natural, therefore, to refine our criterion for equilibrium so that it can be used conveniently to discuss these situations. We can do this by first defining a new function called the **Gibbs free energy** G:[1]

$$G = H - TS \qquad (3\text{-}9)$$

Since G depends only on state functions, it must itself be a function of state with the units of energy. In order to learn the properties of the free energy, we first find that for any infinitesimal change

$$dG = dH - T\,dS - S\,dT$$

Since many chemical processes are isothermal, we restrict ourselves to constant temperature and write

$$dG = dH - T\,dS \qquad \text{(constant } T) \quad (3\text{-}10)$$

or

$$\int_1^2 dG = \Delta G = \Delta H - T\,\Delta S \quad \text{(constant } T)$$

Thus in any isothermal process the change in free energy is very simply related to the changes in enthalpy and entropy of the system.

In order to see how free energy can be used as a criterion for equilibrium, we must introduce the definition of enthalpy into Eq. (3–10):

$$dH = dE + d(PV) = dq - dw + P\,dV + V\,dP$$
$$dG = dq - dw + P\,dV + V\,dP - T\,dS \qquad \text{(constant } T)$$

At this point, a further restriction enters. We consider systems which can do work only by pressure-volume expansions; consequently, for an infinitesimal process

---

[1] Formerly the Gibbs free energy was denoted by F; this symbol will be found in many of the textbooks in the list of supplementary readings.

$$dw = P \, dV$$
$$dG = dq + V \, dP - T \, dS \qquad \text{(constant } T)$$

Most chemical reactions are done in vessels open to the constant pressure of the atomsphere. In such a situation $dP = 0$, and

$$dG = dq - T \, dS \qquad \text{(constant } T, P)$$

Now we introduce the definition of entropy:

$$T \, dS = dq_{\text{rev}} \qquad \text{(constant } T, P)$$

and so

$$dG = dq - dq_{\text{rev}} \qquad \text{(constant } T, P)$$

For a reversible process, $q = q_{\text{rev}}$, and

$$\mathbf{dG = 0}$$
$$\mathbf{\Delta G = 0}$$
(reversible process, constant T, P)

Alternatively, consider an irreversible process. We have shown earlier that

$$q_{\text{irrev}} < q_{\text{rev}}$$
$$dq_{\text{irrev}} < dq_{\text{rev}}$$

Therefore, for an irreversible process

$$\mathbf{dG < 0}$$
$$\mathbf{\Delta G < 0}$$
(irreversible process, constant T, P)

These results say that in a system at constant temperature and pressure G may decrease (irreversible process) or stay constant (reversible process), but never increase. This must mean that when such a system reaches equilibrium, its free energy is a minimum. That is, a system not in equilibrium will tend to change irreversibly, thereby lowering its free energy until no further change is possible. At this point its free energy is a minimum, and the system is at equilibrium. Notice carefully that the conditions under which this argument applies are exactly those which are commonly used in chemical processes: constant temperature and pressure. Furthermore, if we use as our criterion of equilibrium a minimum value of the free energy, we have only to consider the behavior of the system alone, and not of the surroundings. For these reasons, the Gibbs free energy provides the most convenient criterion for equilibrium in chemical systems.

The conclusion that systems tend to change in a way which lowers their free energy is consistent with our physical intuition concerning the thermodynamic functions which determine the free-energy change in an isothermal process:

$$\Delta G = \Delta H - T \Delta S$$

It is common experience that mechanical systems tend to move toward states of low energy. We see that the more negative ΔH is, the more negative ΔG is. Thus, processes in which the system goes from a high- to a low-energy state tend to proceed spontaneously. Furthermore, we see that the more positive ΔS, the more negative ΔG. Therefore, a process in which the system changes from a state of low chaos or probability to one of higher probability also tends to proceed spontaneously, as we would expect. We can say that there are two driving forces which govern the behavior of systems at constant temperature and pressure: the tendency toward minimum enthalpy and the tendency toward maximum entropy. Free energy is the function which assesses both of these tendencies simultaneously.

In some processes ΔH is negative and ΔS is positive. The decomposition of nitrous oxide into its elements is an example of this situation.

$$N_2O(g) \rightarrow N_2(g) + \tfrac{1}{2}O_2(g)$$

$$\Delta H = -19.5 \text{ kcal}$$

$$\Delta S = +18 \text{ cal/deg}$$

$$\Delta G = \Delta H - T \Delta S = -24.9 \text{ kcal} \quad (T = 298°K)$$

Since ΔH is negative and ΔS is positive, ΔG is negative and the decomposition should proceed spontaneously. Although this thermodynamic argument shows that nitrous oxide is unstable, the gas shows no obvious tendency to revert to its elements simply because the *rate* of decomposition at room temperature is small. At elevated temperatures or under catalytic conditions, the reaction occurs as predicted. Thermodynamics can only predict the eventual course of a reaction and the position of equilibrium; it cannot say how fast equilibrium will be attained.

Processes in which ΔH and ΔS have the same sign are more common. For example, in the vaporization of water at temperatures near its normal boiling point we find

$$H_2O(l) \rightarrow H_2O(g, \text{ 1 atm})$$
$$\Delta H = 9590 \text{ cal/mole}$$
$$\Delta S = 25.7 \text{ cal/mole-deg}$$

In this case ΔG will be negative and the evaporation to form vapor *at 1 atm pressure* will be spontaneous if

$$T \, \Delta S > \Delta H$$

For this to be true the temperature must satisfy the inequality

$$T > \frac{\Delta H}{\Delta S}$$

$$T > \frac{9590}{25.7}$$

$$T > 373$$

That is, ΔG is negative and the evaporation to form vapor at 1 atm pressure is spontaneous *if the temperature is above the normal boiling point of water*. On the other hand, ΔG will be positive if

$$T \, \Delta S < \Delta H$$

This will be true if

$$T < \frac{\Delta H}{\Delta S}$$

$$T < 373$$

As long as the temperature is below the boiling point, the formation of vapor at *1-atm pressure* is not a spontaneous process. However, for the *reverse process*

$$H_2O(g, \text{ 1 atm}) \rightarrow H_2O(l) \qquad (T < 373)$$

ΔG is negative and the condensation of the supersaturated vapor proceeds spontaneously.

The boiling temperature is the temperature at which the equilibrium vapor pressure is 1 atm. At 373°K we have

$$T \, \Delta S = \Delta H \qquad \Delta G = 0$$

The transfer of 1 mole of water from the liquid to the vapor at 1 atm pressure at 373°K involves no change in free energy and is therefore a reversible process.

This example shows that when ΔH and ΔS have the same sign, ΔG may be either positive or negative, depending on the temperature. Temperature is a weighting factor which determines the relative importance of enthalpy and entropy. At high temperatures the entropy is the driving force which dominates the behavior of systems, while at low temperatures ΔH and the tendency toward low energy are more important.

3-6 FREE ENERGY
AND THE EQUILIBRIUM CONSTANT

In our discussion of the consequences of the first law of thermodynamics we found that it was very useful to assign absolute values of the enthalpy to each chemical compound. We can assign absolute free energies in the same manner. First, we assign every *element* a free energy of zero when it is in its standard state. Then by methods to be discussed later, we measure the free energy change which accompanies the formation of a compound in its standard state from its elements. This free energy change is the standard free energy of formation ΔG_f° of the compound. Table 3–2 gives the free energy of formation of several compounds. With such a table at hand, we can calculate the standard free energy change ΔG° for any reaction. The procedure, as our experience with ΔH_f° might suggest, is

$$\Delta G^\circ = \Sigma \, \Delta G_f^\circ(\text{products}) - \Sigma \, \Delta G_f^\circ(\text{reactants}) \qquad (3\text{-}11)$$

The standard free energy change is a fixed constant which measures the difference in free energy between two very definite states: the standard state of the reactants and the standard state of the products. Thus ΔG° depends only on the nature of the

Table 3-2
Free Energy of Formation
ΔG_f°, kcal/mole, at 298°K

Gases		Solids	
H_2O	-54.64	AgCl	-26.22
H_2O_2	-24.7	AgBr	-22.39
O_3	39.06	AgI	-15.81
HCl	-22.77	BaO	-126.3
HBr	-12.72	$BaSO_4$	-350.2
HI	0.31	$BaCO_3$	-272.2
SO_2	-71.79	CaO	-144.4
SO_3	-88.52	$CaCO_3$	-269.8
H_2S	-7.89	$Ca(OH)_2$	-214.3
N_2O	24.9	SiO_2	-192.4
NO	20.72	Fe_2O_3	-177.1
NO_2	12.39	Al_2O_3	-376.8
NH_3	-3.97	CuO	-30.4
CO	-32.81	Cu_2O	-34.98
CO_2	-94.26	ZnO	-76.05

Organic compounds
Gases

Methane, CH_4	-12.14	Ethylene, C_2H_4	16.28
Ethane, C_2H_6	-7.86	Acetylene, C_2H_2	50.00
Propane, C_3H_8	-5.61	1-Butene, C_4H_8	17.09
n-Butane, C_4H_{10}	-3.75	cis-2-Butene, C_4H_8	15.74
Isobutane, C_4H_{10}	-4.3	$trans$-2-Butene, C_4H_8	15.05
n-Pentane, C_5H_{12}	-2.0	Isobutene, C_4H_8	13.88
Isopentane, C_5H_{12}	-3.5	1,3-Butadiene, C_4H_6	36.01
Neopentane, C_5H_{12}	-3.6	Methyl chloride, CH_3Cl	-14.0

Liquids

Methanol, CH_3OH	-39.73	Benzene, C_6H_6	29.76
Ethanol, C_2H_5OH	-41.77	Chloroform, $CHCl_3$	-17.1
Acetic acid, CH_3COOH	-93.8	Carbon tetrachloride, CCl_4	-16.4

Aqueous ions

H^+	0.0	OH^-	-37.59
Na^+	-62.59	Cl^-	-31.35
K^+	-67.47	Br^-	-24.57

Table 3–2 (*continued*)

Ag$^+$	18.43	I$^-$	−12.35
Ba^{2+}	−134.0	HS$^-$	3.01
Ca^{2+}	−132.18	S^{2-}	20.0
Cu^{2+}	15.53	SO$_4^{2-}$	−177.34
Zn^{2+}	−35.18	SO$_3^{2-}$	−126.2

Gaseous atoms

H	48.57	I	16.77
F	14.2	C	160.84
Cl	25.19	N	81.47
Br	19.69	O	54.99

reactants and the products and on the temperature. The sign of the standard free energy change tells us whether the process

reactants(1 atm, 1 M, 298°K)
$\rightarrow$ products(1 atm, 1 M, 298°K)

will occur spontaneously. If $\Delta G°$ has a negative sign, this means reaction will be sufficiently complete that reactants *in their standard states* will be converted spontaneously to products *in their standard states*. If $\Delta G°$ is a positive number, then the reactants in their standard states may react to form *some product*, but they will not react sufficiently completely to form products in their standard states. In other words, a positive $\Delta G°$ means that conversion of reactants in their standard states to products in their standard states is not a spontaneous process. Therefore, the sign of the standard free energy change for a reaction gives some indication of how completely reactants will be converted to products. A more detailed description of the concentration conditions which exist when a reaction reaches equilibrium is provided by the equilibrium constant. The standard free energy change and the equilibrium constant of a reaction are mathematically related, as we shall now see.

Consider a hypothetical reaction between ideal gases

$$aA + bB \rightarrow cC + dD$$

We shall develop an expression for the free energy change which occurs when the reactants at arbitrary pressures P_A, P_B, are converted to products at arbitrary pressures P_C, P_D. We shall then use this expression to determine the relation between $\Delta G°$ and the equilibrium constant. To do this, we refer to the definition of free energy and write

$$dG = dH - T \, dS - S \, dT$$
$$= dq + V \, dP - T \, dS - S \, dT$$
$$\boldsymbol{dG = V \, dP - S \, dT} \qquad \qquad \textbf{(3–12)}$$

This very important equation shows how free energy depends on pressure and temperature. For an isothermal pressure change we have

$$dG = V \, dP$$
$$\int_{G°}^{G} dG = \int_{P_0}^{P} V \, dP \qquad \qquad \text{(constant } T\text{)}$$

Since we are discussing only ideal gases,

$$\bar{G} - \bar{G}° = \int_{P_0}^{P} \frac{RT}{P} \, dP = RT \ln \frac{P}{P_0}$$

In this equation, $\bar{G}$ is the free energy of *one mole* of an ideal gas at an arbitrary pressure P and $\bar{G}°$ is the standard free energy of one mole of the gas. This means that $P_0 = 1$ atm, the standard pressure, and we write

$$\bar{G} = \bar{G}° + RT \ln \frac{P}{1} = \bar{G}° + RT \ln P \qquad \qquad \textbf{(3–13)}$$

or for n moles

$$n\bar{G} = n\bar{G}° + nRT \ln P \qquad \qquad \textbf{(3–14)}$$

We can write similar expressions for each reactant and product of our chemical reaction. To find ΔG, the free energy change, we write

$$\Delta G = \Sigma \, G(\text{products}) - \Sigma \, G(\text{reactants})$$
$$= c\bar{G}_C + d\bar{G}_D - a\bar{G}_A - b\bar{G}_B$$

Now we substitute an expression of the form of Eq. (3–14) for the free energy of each reactant and product and find

$$\Delta G = c\bar{G}_C^\circ + d\bar{G}_D^\circ - a\bar{G}_A^\circ - b\bar{G}_B^\circ + cRT \ln P_C$$
$$+ dRT \ln P_D - aRT \ln P_A - bRT \ln P_B \qquad (3\text{–}15)$$

But

$$c\bar{G}_C^\circ + d\bar{G}_D^\circ - a\bar{G}_A^\circ - b\bar{G}_B^\circ = \Delta G^\circ$$

That is, the standard free energy change of the reaction is just the difference between the standard free energies of products and reactants. Therefore, we can write Eq. (3–15) as

$$\Delta G = \Delta G^\circ + RT \ln \frac{(P_C)^c (P_D)^d}{(P_A)^a (P_B)^b} \qquad (3\text{–}16)$$

This general expression gives the free energy change which accompanies the conversion of reactants at arbitrary pressures P_A, P_B to products at arbitrary pressures P_C, P_D. The quotient composed of the pressures of reactants and products has the *form* of an equilibrium constant; each pressure is raised to a power which is the stoichiometric coefficient of the compound in the chemical reaction. However, this concentration quotient is *not equal* to the equilibrium constant unless the pressures P_A, P_B, P_C, P_D are those which exist at chemical equilibrium. When a reaction does reach equilibrium, the free energy of the products is the same as that of the reactants.

If this latter statement were not true, some reaction, either forward or reverse, which would lower the free energy of the system could take place. If the free energy of reactants and products is the same, the reaction has no tendency to proceed in either direction. Therefore, if we take reactants *in their equilibrium concentrations* and convert them to products *in their equilibrium concentrations*, then

$$\Delta G = 0$$

and

$$0 = \Delta G^\circ + RT \ln \left[\frac{(P_C)^c (P_D)^d}{(P_A)^a (P_B)^b} \right]_{\text{equil}}$$

$$0 = \Delta G^\circ + RT \ln K$$

$$\Delta G^\circ = -RT \ln K \qquad (3\text{–}17)$$

The equilibrium constant K and the standard free energy change are related in a remarkably simple way. Remember that we emphasized that $\Delta G°$ is a fixed constant that depends, at a fixed temperature, *only on the nature of the reactants and* products. Equation (3–17) tells us that since $\Delta G°$ is a fixed constant, there is a constant K which at a given temperature must also depend only on the nature of reactants and products. Our derivation then shows that the equilibrium constant is a true constant and has the functional form to which we are accustomed. This conclusion rests on the well-supported fact that G is a function of state.

One more remark should be made about Eqs. (3–16) and (3–17). When we write

$$\bar{G} = \bar{G}° + RT \ln P$$

are we taking the logarithm of a number which has units? Not really, since this equation is actually

$$\bar{G} = \bar{G}° + RT \ln \frac{P}{P_0} = \bar{G}° + RT \ln \frac{P}{1}$$

where the 1 in the denominator of the pressure term also has units. Therefore the units of pressure cancel, and we really always take the logarithm of a *dimensionless number* both here and in the expression involving the equilibrium constant.

We have obtained the relation between the equilibrium constant and the standard free energy for the special case of reaction between ideal gases. Can this result be generalized to include reactions which take place in solution? Experience shows the free energy of any substance, whether in the gaseous or condensed phase, can be expressed by an equation of the form of Eq. (3–14). For dilute solutions the relation between solute concentration C and free energy is

$$\bar{G} = \bar{G}° + RT \ln C \qquad\qquad (3\text{–}18)$$

where $\bar{G}°$ is the free energy of the solute when its concentration is unity. Since this is the case, the relationship in Eq. (3–17) between $\Delta G°$ and K is preserved for reactions which take place in dilute solution, with the equilibrium constant expressed in terms of concentrations.

Let us examine the consequences of Eq. (3–17). The standard free energy change $\Delta G°$ is a number generally different from zero. If $\Delta G°$ is a negative number, we concluded earlier, the conversion of reactants *in their standard states* to products *in their standard states* is spontaneous. Equation (3–17) shows that if $\Delta G°$ is negative, the equilibrium constant must be greater than unity. Furthermore, the more negative $\Delta G°$ is, the larger K is.

If $\Delta G°$ is positive, we can conclude that the conversion of reactants in their standard states to products in their standard states is not spontaneous. Does this mean that no reaction takes place? Not at all! It merely means that the equilibrium constant for the reaction is less than unity. The more positive $\Delta G°$ is, the smaller K is. The special case in which $\Delta G° = 0$ corresponds to an equilibrium constant of unity.

EXAMPLE 3–3

For the decomposition of gaseous hydrogen peroxide according to the reaction

$$H_2O_2(g) = H_2O(g) + \tfrac{1}{2}O_2(g)$$

$\Delta G° = -29.9$ kcal. What is the value of the equilibrium constant of this reaction at 298°K?

From Eq. (3–17), $\Delta G° = -RT \ln K$, so

$$K = 10^{-\Delta G°/2.3RT} = 10^{29,900/1370} = 10^{21.8}$$

Thus a negative $\Delta G°$ corresponds to a large equilibrium constant. Since the standard state for gases is 1 atm, the concentration units to be used with the equilibrium constant are atmospheres.

EXAMPLE 3–4

When gaseous ozone, O_3, is formed from molecular oxygen by the reaction

$$\tfrac{3}{2}O_2(g) = O_3(g)$$

the standard free energy change is 39.1 kcal. What is the equilibrium constant of this reaction at 298°K?

By Eq. (3–17)

$$K = 10^{-\Delta G°/2.3RT} = 10^{-39,100/1370} = 10^{-28.4}$$

Thus a large positive standard free energy change corresponds to a small equilibrium constant.

EXAMPLE 3–5

The equilibrium pressure of water vapor over a mixture of $CuSO_4$ and $CuSO_4 \cdot H_2O$ is 4.5 mm at 323°K. Compute ΔG for the following processes:

(a) $CuSO_4 \cdot H_2O(s) = CuSO_4(s) + H_2O(g, P = 4.5 \text{ mm})$

(b) $CuSO_4 \cdot H_2O(s) = CuSO_4(s) + H_2O(g, P = 760 \text{ mm})$

For the first process, the initial and final states are just those that are in equilibrium at 323°K; therefore, for this process, $\Delta G = 0$. In the second process, the initial and final states are the standard states of reactant and product at 323°K, so

$$\Delta G = \Delta G^\circ = -RT \ln K$$

The equilibrium constant must be expressed in atmospheres to calculate ΔG°, so

$$\Delta G^\circ = -(2.30)(1.98)(323) \, [\log (4.5/760)]$$
$$\Delta G^\circ = +3290 \text{ cal}$$

Example 3–5 illustrates the important difference between ΔG and ΔG°. When two states are in equilibrium with each other, the difference ΔG in their free energies is equal to zero. The difference ΔG° in the free energy of their standard states is in general not equal to zero.

3–7   DETERMINATION OF FREE ENERGY CHANGES

In the preceding section we learned how to calculate the equilibrium constant of a reaction from the standard free energy change ΔG°. Now we shall explore procedures for determining ΔG° experimentally. The most obvious method of finding ΔG° involves the relation between free energy, enthalpy, and entropy:

$$\Delta G^\circ = \Delta H^\circ - T \Delta S^\circ \qquad (3\text{–}19)$$

This equation is just the definition of the free energy change in an isothermal process applied to the special case when the initial and final states of the system are the standard states of the reactants and products. The terms in Eq. (3–19) can be evaluated quite easily:

$$\Delta H^\circ = \Sigma \, \Delta H_f^\circ(\text{products}) - \Sigma \, \Delta H_f^\circ(\text{reactants})$$

That is, the standard enthalpy change is the difference between the enthalpies of formation of the products and reactants, all in their standard states. Similarly,

$$\Delta S° = \Sigma\, S°(\text{products}) - \Sigma\, S°(\text{reactants})$$

Therefore, access to a table of enthalpies of formation and of absolute entropies permits us to evaluate $\Delta H°$ and $\Delta S°$ for any any reaction, as long as $\Delta H_f°$ and $S°$ have been tabulated for each reactant and product. Since the standard free energies and entropies of compounds are generally reported at 298°K, insertion of this temperature along with $\Delta H°$ and $\Delta S°$ into Eq. (3–19) gives $\Delta G°$ at 298°K.

As an example we consider the reaction

$$N_2 + O_2 = 2NO$$

From Table 2–2 we find the enthalpy of formation of 1 mole of gaseous NO is 21.60 kcal. The enthalpies of formation of nitrogen and oxygen are zero, since they are elements in their standard states. Therefore,

$$\Delta H° = 2 \times \Delta H_f°(\text{NO}) = 43.20 \text{ kcal}$$

To determine $\Delta S°$, we consult Table 3–1 and find

$$S°(\text{NO}) = 50.34 \text{ cal/mole-deg}$$
$$S°(\text{N}_2) = 45.77$$
$$S°(\text{O}_2) = 49.01$$
$$\Delta S° = 2S°(\text{NO}) - S°(\text{N}_2) - S°(\text{O}_2) = 5.90 \text{ cal/deg}$$

By combining $\Delta H°$ and $\Delta S°$, we obtain

$$\Delta G° = 43,200 - 298 \times 5.90$$
$$\Delta G° = 41,400 \text{ cal}$$

Since $\Delta G°$ is positive, we expect the equilibrium constant for the formation of nitric oxide at 298°K to be less than unity.

$$\Delta G° = -RT \ln K = -2.3RT \log K$$
$$\log K = -\Delta G°/2.3RT$$
$$K = 10^{-\Delta G°/2.3RT}$$
$$K = 10^{-\Delta G°/1370}$$
$$K = 10^{-30.2} = 6 \times 10^{-31}$$

The equilibrium constant is indeed very small, principally because the reaction is so endothermic. In fact, this equilibrium constant is so small at room temperature that there would be absolutely no hope of determining it by direct chemical analysis of an equilibrium mixture of N_2, O_2, and NO. However, measurements of ΔH_f° and S° of the individual reactants and products allow us to calculate K, irrespective of its magnitude, with no difficulty. Notice carefully that S° for each reagent can be obtained by making thermal measurements (heat capacities) on the *pure compounds*. Determination of ΔH_f° involves a thermal measurement on one convenient reaction of a compound. The calculation of chemical equilibrium constants from thermal measurements on individual reagents is one of the great triumphs of chemical thermodynamics.

When the equilibrium constant is very large or very small, calculation using free energies of formation is often the only way to obtain the exact value of K. On the other hand, when K is near unity and the reaction is relatively uncomplicated, it is often useful to reverse the normal procedure and obtain free energies from directly measured equilibrium constants. For example, ΔG_f° of gaseous hydrogen iodide might be found from its measured enthalpy and entropy of formation, but since the equilibrium constant of the reaction

$$H_2(g) + I_2(g) = 2HI(g)$$

is near unity and relatively easy to measure, direct evaluation of K, and thus of ΔG°, leads to a reliable value of $\Delta G_f^\circ(HI)$. This procedure involves less work than measurement of ΔH_f° and S° for the three substances involved in the reaction. The value of $\Delta G_f^\circ(HI)$ obtained from the measured equilibrium constant may be used to help calculate the value of the equilibrium constant for any of the many other reactions which involve HI. Thus, by use of the free energy concept, measurement of one equilibrium constant can help to determine many others.

Although calculation from thermal measurements of ΔH° and ΔS° and use of measured equilibrium constants are two important methods of obtaining standard free energy changes, there is still a third technique which we shall treat in the next section.

3-8 THE ELECTROCHEMICAL CELL

Some chemical reactions can be run in electrochemical cells, and it is possible to associate a voltage or cell potential with the reaction. We shall see that this cell potential is very simply related to the free energy change for the reaction. Therefore, measurement of cell potentials is often a very convenient way to obtain free energy changes and consequently equilibrium constants. To derive the connection between free energy and cell potentials, we again return to the definition of free energy and write

$$dG = dq - dw + P \, dV + V \, dP - S \, dT - T \, dS$$
$$dG = dq - dw + P \, dV - T \, dS \qquad \text{(constant } T, P)$$

We have already argued that the work done by the system in a reversible change is the maximum work the system can do. Therefore, we write

$$dG = dq_{\text{rev}} - dw_{\text{max}} + P \, dV - T \, dS \quad \text{(constant } T, P, \text{ reversible process)}$$
$$dq_{\text{rev}} = T \, dS$$
$$dG = -dw_{\text{max}} + P \, dV$$
$$\Delta G = -w_{\text{max}} + P \, \Delta V$$

Until now we have considered systems that were capable of doing work only by a pressure-volume expansion. However, when a chemical reaction occurs in an electrochemical cell, the system may do work not only by a pressure-volume expansion but also by transferring electrical charge through an external circuit. Furthermore, it is relatively easy to make these electrochemical cells operate reversibly, since all we need do is arrange a circuit that draws no more than an infinitesimal current from the cell. Consequently, we can write

$$w_{\text{rev}} = w_{\text{max}} = w_{\text{elec}} + w_{PV}$$

and

$$\Delta G = w_{\text{elec}} - w_{PV} + P \, \Delta V \quad \text{(constant } T, P, \text{ reversible process)}$$
$$\Delta G = -w_{\text{elec}} \qquad \qquad \text{(3-20)}$$

That is, the free energy difference between two states is equal to the negative of the electrical work that is done by the system as it goes reversibly from one state to the other. If the free energy of the system decreases, ΔG is negative and the system does electrical work on its surroundings. If ΔG is positive, the contemplated change can be effected if the surroundings do electrical work on the system, as in an electrolysis.

Now we must relate the electrical work to the voltage of the electrochemical cell. The voltage difference between any two points is defined as the work done in transferring a unit charge between them. For a reaction occurring in an electrochemical cell whose electrodes differ in potential by $\Delta \mathcal{E}$, the work done when amount of charge $n\mathcal{F}$ is transferred is $n\mathcal{F} \Delta \mathcal{E}$, so

$$\Delta G = -n\mathcal{F} \Delta \mathcal{E} \qquad (3\text{-}21)$$

Here $\mathcal{F}$ is the Faraday,[1] the charge on 1 mole of electrons, and n is the number of moles of charge transferred.

Experimentally observed cell voltages depend on the nature of the reactants and products and also on their concentration and temperature. To compare cell voltages on a common basis, we define a standard voltage $\Delta \mathcal{E}°$ which is the measured voltage of a cell when the reactants and products are in their standard states at 298°K. The free energy associated with the standard cell potential must then be the standard free energy change for the reaction, $\Delta G°$. That is,

$$\Delta G° = -n\mathcal{F} \Delta \mathcal{E}° \qquad (3\text{-}22)$$

This equation shows that standard free energy changes, and therefore equilibrium constants, can be obtained from measurements of cell potentials. For example, consider the reaction

$$Zn + Cu^{2+}(aq) \rightarrow Cu + Zn^{2+}(aq)$$

The voltage of a cell in which a zinc metal electrode dips into 1-M Zn^{2+} and a copper electrode dips into 1-M Cu^{2+} is 1.100 volts. The corresponding free energy change is

---

[1] The most recent determination of $\mathcal{F}$ gives 96,486 coulombs, based on the new C^{12} atomic weight scale.

$$\Delta G° = -n\mathfrak{F}\,\Delta \mathcal{E}° = -(2)(96,486)(1.100) = -212,300 \text{ joules}$$
$$= -(2)(23,061)(1.100) = -50,700 \text{ cal}$$

Since free energy is usually expressed in calories, it is useful to remember that 23,060 is the numerical value of the Faraday[1] which, when multiplied by voltage, gives energy in units of calories.

To find the equilibrium constant of this reaction, we write

$$\Delta G° = -RT \ln K$$

or

$$K = 10^{-\Delta G°/2.3RT}$$
$$= 10^{+37.3} = 2 \times 10^{37}$$

This is a remarkably large number, which indicates that very little Cu^{2+} could be found in an equilibrium mixture of Cu, Cu^{2+}, Zn^{2+}, and Zn. Again, there is no possibility of evaluating this equilibrium constant by direct chemical analysis of the equilibrium mixture, since the concentration of Cu^{2+} would be immeasurably small. The determination of the standard cell potential is done with all reagents at $1\ M$ concentration, which is extremely convenient. The calculation of equilibrium constants from cell voltage measurements is another example of how thermodynamics shows us how to use the results of convenient experiments to calculate quantities which would be difficult or impossible to obtain by direct measurement.

Let us pursue the consequences of the relation between free energy and cell potential. According to Eq. (3–16), the free energy change for any process involving arbitrary concentrations of reagents

$$a\mathrm{A} + b\mathrm{B} = c\mathrm{C} + d\mathrm{D}$$

can be expressed as

$$\Delta G = \Delta G° + RT \ln \frac{(\mathrm{C})^c(\mathrm{D})^d}{(\mathrm{A})^a(\mathrm{B})^b}$$

---

[1] This value is derived by writing
$$\mathfrak{F} = 96,486 \text{ coulombs} = 96,486 \text{ volt-coulombs/volt}$$
$$= 96,486 \text{ joules/volt}$$
$$1 \text{ cal} = 4.1840 \text{ joules}$$
$$\mathfrak{F} = 96,486/4.1840 = 23,060 \text{ cal/volt}$$

where the concentration terms may be in atmospheres, molarity, or any other unit which proves convenient. Substituting the relation between free energy and cell potential gives

$$-n\mathfrak{F} \, \Delta\mathcal{E} = -n\mathfrak{F} \, \Delta\mathcal{E}^\circ + RT \ln \frac{(C)^c(D)^d}{(A)^a(B)^b}$$

$$\Delta\mathcal{E} = \Delta\mathcal{E}^\circ - \frac{RT}{n\mathfrak{F}} \ln \frac{(C)^c(D)^d}{(A)^a(B)^b}$$

If we make the further substitutions

$$R = 1.98 \text{ cal/mole-deg}$$
$$T = 298°\text{K}$$
$$\mathfrak{F} = 23{,}100 \text{ cal/volt}$$
$$\ln_e = 2.3 \log_{10}$$

we obtain

$$\Delta\mathcal{E} = \Delta\mathcal{E}^\circ - \frac{0.059}{n} \log \frac{(C)^c(D)^d}{(A)^a(B)^b} \qquad (3\text{-}23)$$

This expression is called the Nernst equation, and it relates the measured cell potential $\Delta\mathcal{E}$ to the standard cell potential $\Delta\mathcal{E}^\circ$ and the existing concentrations of the reactants and products of the cell reaction. It is supported by a wealth of experimental verification. In the special situation in which all concentrations are unity, the logarithmic term becomes zero, and $\Delta\mathcal{E} = \Delta\mathcal{E}^\circ$.

The other important special case occurs when all concentrations are equilibrium concentrations. Then no net reaction occurs, no charge is transferred spontaneously, and therefore the electrodes must be at the same potential. Thus $\Delta\mathcal{E} = 0$ and

$$0 = \Delta\mathcal{E}^\circ - \frac{0.059}{n} \log K$$

$$K = 10^{n \, \Delta\mathcal{E}^\circ/0.059}$$

We could have obtained this expression directly from Eqs. (3-17) and (3-22).

The operation of the electrochemical cell provides an experimental demonstration that a reaction proceeds until the free energy of reactants and products is the same. Owing to the relation between cell voltage and free energy difference, a voltmeter is essentially a free energy difference meter. While the cell reaction proceeds, the voltage difference between the electrodes falls as the free energies of products and reactants approach each other. When the electrodes reach the same voltage, charge transfer ceases and the reaction is at equilibrium.

3–9   TEMPERATURE DEPENDENCE OF EQUILIBRIA

Besides providing us with convenient indirect methods of evaluating equilibrium constants, thermodynamics shows how equilibria are affected by changes in temperature. In addition to depending on the nature of products and reactants, $\Delta G°$, and hence K, is a function of temperature. Let us combine the fundamental equations

$$\Delta G° = -RT \ln K$$
$$\Delta G° = \Delta H° - T \Delta S°$$

to give

$$\ln K = -\frac{\Delta H°}{RT} + \frac{\Delta S°}{R} \qquad (3\text{–}24)$$

Equation (3–24) tells us that *if* $\Delta H°$ and $\Delta S°$ are constants *independent of temperature*, the logarithm of the equilibrium constant is a simple linear function of the reciprocal temperature. Now, Eq. (2–17) assures us that $\Delta H°$ will indeed be independent of temperature if $\Delta C_P = 0$, that is, if the heat capacities of reactants and products are the same. But what of the temperature dependence of $\Delta S°$? According to Eq. (3–4), the entropy of any substance is given by the expression

$$S_T° - S_{298}° = \int_{298}^{T} C_P \frac{dT}{T}$$

If we write such an equation for each reactant and product and then take their difference to find $\Delta S_T°$, we have

$$\Delta S_T° = \Delta S_{298}° + \int_{298}^{T} \Delta C_P \frac{dT}{T} \qquad (3\text{–}25)$$

where once again ΔC_P is the difference in the heat capacities of reactants and products.

Equation (3–25) shows that if ΔC_P is zero, $\Delta S°$ is independent of temperature. Subject to this condition then, the equation

$$\ln K = -\frac{\Delta H°}{RT} + \frac{\Delta S°}{R}$$

says that $\ln K$ is a linear function of $1/T$. This can be expressed more compactly by differentiating the equation with respect to $1/T$. The result is

$$\frac{d(\ln K)}{d(1/T)} = -\frac{\Delta H°}{R} \qquad (3\text{–}26)$$

or, since $d(1/T) = -dT/T^2$,

$$\frac{d(\ln K)}{dT} = \frac{\Delta H°}{RT^2} \qquad (3\text{–}27)$$

These equations show that it is the $\Delta H°$ of a reaction that determines the effect of temperature on its equilibrium constant. If $\Delta H°$ is positive, then Eq. (3–27) shows that $d(\ln K)$ is positive if dT is positive. Thus for an endothermic reaction, K increases as T increases. If $\Delta H°$ is negative, K decreases with increasing temperature. These conclusions are consistent with Le Chatelier's principle, which states that increasing the temperature of an equilibrium reaction mixture causes the reaction to proceed in the direction which absorbs heat. If the reaction is endothermic, then increasing the temperature increases the product concentration, and this can be done only by increasing the equilibrium constant. On the other hand, the equilibrium constants of exothermic reactions must decrease as temperature increases.

Equation (3–26) can be written in the form

$$d(\ln K) = -\frac{\Delta H°}{R} d\left(\frac{1}{T}\right)$$

and integrated between two temperatures T_1 and T_2, at which the equilibrium constant has the values K_1 and K_2, respectively. Assuming that $\Delta H°$ is independent of temperature, we have

$$\int_{K_1}^{K_2} d(\ln K) = -\frac{\Delta H^\circ}{R} \int_{T_1}^{T_2} d\left(\frac{1}{T}\right) \qquad (3\text{-}28)$$

$$\ln \frac{K_2}{K_1} = -\frac{\Delta H^\circ}{R}\left(\frac{1}{T_2} - \frac{1}{T_1}\right) \qquad (3\text{-}29)$$

Equation (3–29) shows that knowledge of the equilibrium constant at one temperature permits us to calculate the equilibrium constant at any other temperature, provided we know the ΔH° of the reaction. Conversely, measurement of K at two different temperatures allows us to calculate ΔH° for the reaction. In fact, Eq. (3–24) indicates that a plot of $\ln K$ as a function of $1/T$ should be a straight line whose slope is $-\Delta H^\circ/R$. Figure 3–6 shows this

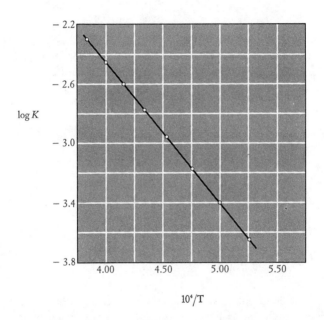

Figure 3–6 The logarithm of the equilibrium constant for the reaction $N_2 + O_2 = 2NO$ as a function of the reciprocal of absolute temperature. (Redrawn from Daniels and Alberty, *Physical Chemistry*, 2d ed., Wiley, New York, 1961.)

type of a graph for the reaction between nitrogen and oxygen to form nitric oxide.

EXAMPLE 3-6

The equilibrium constant for the reaction

$$2NO_2(g) = N_2O_4(g)$$

is 8.8 at 298°K, and $\Delta H° = -13.9$ kcal. What is the value of the equilibrium constant at 273°K? In Eq. (3–29) we set $K_1 = 8.8$, $T_1 = 298°K$, and $T_2 = 273°K$ and get

$$\ln \frac{K_2}{K_1} = - \frac{\Delta H°}{R} \left(\frac{1}{T_2} - \frac{1}{T_1} \right)$$

$$2.30 \log \frac{K_2}{8.8} = \frac{13,900}{1.98} \left(\frac{1}{273} - \frac{1}{298} \right)$$

$$K_2 = 76$$

Since the reaction is exothermic, we expect the equilibrium constant to be larger at the lower temperature, as is found.

EXAMPLE 3-7

The equilibrium constant for the reaction

$$2HI(g) = H_2(g) + I_2(g)$$

is 2.18×10^{-2} at 764°K, and 1.64×10^{-2} at 667°K. Find $\Delta H°$ for the reaction.

Again we use Eq. (3–29). Substitution of the data gives us a numerical expression for $\Delta H°$:

$$\ln \left(\frac{1.64 \times 10^{-2}}{2.18 \times 10^{-2}} \right) = - \frac{\Delta H°}{1.98} \left(\frac{1}{667} - \frac{1}{764} \right)$$

$$\Delta H° = 2950 \text{ cal}$$

Thus, measurement of K at two different temperatures gives ΔH for the reaction.

In this chapter we have seen that the principles of thermodynamics can be used to predict or understand the direction and extent of spontaneous change. In totally isolated systems where the energy and enthalpy are constant, the only changes that occur spontaneously are those in which the entropy of the system increases. The eventual position of equilibrium for such systems is the state of maximum entropy. For systems maintained at constant temperature and pressure the only spontaneous processes are those in which the free energy of the system decreases. Equilibrium is reached when the free energy is a minimum. These ideas lead to a connection between the equilibrium constant of a reaction and the properties of the reactants and products. Thus we can predict equilibrium constants by using data obtained from measurements of the properties of individual compounds and elements. In the following chapter we shall put this technique to use by studying its application to several types of chemical and physical phenomena.

PROBLEMS

1. Find the algebraic expressions for the ΔS of steps a, b, and c of Fig. 2–11. Use these expressions to show that ΔS for the gas is independent of the path taken between states.

2. By following the procedure used to derive Eq. (3–2), find the expression for the entropy change which accompanies the isothermal expansion of 1 mole of a gas whose equation of state is $P(V - b) = RT$. What is the value of ΔS for 1 mole of gas if $V_1 = 0.224$ liter, $V_2 = 22.40$ liters, and $b = 0.040$ liter? Is this greater or less than the corresponding entropy change for an ideal gas?

3. Calculate the entropy of vaporization of the following liquids.

	T_b, °K	ΔH_{vap}, kcal		T_b, °K	ΔH_{vap}, kcal
Cl_2	269	4.87	$PbCl_2$	1145	24.8
C_6H_6	353	7.35	H_2O	373	9.72
$CHCl_3$	334	7.02	C_2H_5OH	351	9.22

By using the fact that, in liquid water and ethanol, the molecules are partially immobilized by hydrogen bonds which link the hydroxyl group of one molecule with that of another, explain why water and ethanol do not obey Trouton's rule.

4. When a mole of water supercooled to $-10°C$ freezes iso-thermally, what is its entropy change? The process as described is irreversible, so in order to calculate ΔS, a reversible path between initial and final states must be found. One such path is

$$H_2O(l), \ -10°C = H_2O(l), \ 0°C$$
$$H_2O(l), \ 0°C = H_2O(s), \ 0°C$$
$$H_2O(s), \ 0°C = H_2O(s), \ -10°C$$

The molar enthalpy of fusion of ice at $0°C$ is 1440 cal, the molar heat capacity of ice is 9.0 cal/mole-deg, and the molar heat capacity of water is 18.0 cal/mole-deg. Use these data to calculate ΔS of the water when it freezes at $-10°C$.

The enthalpy of fusion of ice at $-10°C$ is 1350 cal/mole. Find the entropy change of the surroundings when 1 mole of water freezes at $-10°C$. What is the total entropy change of the universe for this process? Is the process irreversible according to the second law of thermodynamics?

5. Despite the fact that we derived Eq. (3–27) under the assumption that $\Delta H°$ is independent of temperature, the use of the equation is not restricted to such cases. To find how $\ln K$ depends on temperature when $\Delta H°$ is not a constant, express $\Delta H°$ at any temperature as

$$\Delta H° = \Delta H_0° + \Delta C_P \, T$$

where $\Delta H_0°$ is a true constant independent of temperature. Substitute this expression for $\Delta H°$ in Eq. (3–27) and integrate between the limits T_1, T_2 and K_1, K_2 as in Eq. (3–28). The result should be similar to Eq. (3–29), but with a correction term which accounts for the temperature dependence of $\Delta H°$.

6. Calculate the entropy changes of the following chemical reactions, all at $298°K$.

$$Ba(s) + \tfrac{1}{2}O_2(g) = BaO(s)$$
$$BaCO_3(s) = BaO(s) + CO_2(g)$$
$$Br_2(g) = 2Br(g)$$
$$H_2(g) + Br_2(l) = 2HBr(g)$$

Explain the sign of each of the entropy changes by qualitatively comparing the freedom of motion, or molecular chaos, which characterizes reactants and products.

7. Calculate the absolute entropy of liquid chlorine at its freezing point, 172.1°K, from the following data:

T,°K	15	20	30	40	50	60	70
C_P, cal/mole-deg	0.89	1.85	3.99	5.73	6.99	8.00	8.68

T,°K	90	110	130	150	170	172
C_P, cal/mole-deg	9.71	10.47	11.29	12.20	13.17	mp

The enthalpy of fusion of chlorine is 1531 cal/mole. The contribution to the entropy of temperatures below 15°K is 0.33 eu. The calculation is eased considerably if C_P is plotted as a function of $\log T$.

8. Metallic aluminum is an effective reagent for the reduction of oxides of some other metals, as in the thermite reaction:

$$2Al(s) + Fe_2O_3(s) = Al_2O_3(s) + 2Fe(s)$$

By using the sign of the standard free energy change as a criterion, decide which of the following oxides cannot be reduced by aluminum: CuO, SiO_2, ZnO, CaO.

9. Find the value of $\Delta G°$ and K at 298°K for the reaction

$$CaCO_3(s) = CaO(s) + CO_2(g)$$

What is the pressure of CO_2 that is in equilibrium with a mixture of $CaCO_3$ and CaO at 298°K? Measurement of the pressure of CO_2 in equilibrium with a $CaO–CaCO_3$ mixture at various temperatures gave the following data:

P_{CO_2}, atm	0.029	1.05	3.9	11.5	28.6
T, °K	973	1173	1273	1373	1473

Find $\Delta H°$ of the reaction by using a graphical procedure, and compare your answer with that found by use of the enthalpies of formation.

10. Small amounts of hydrogen can safely be removed from gaseous mixtures by reaction with cupric oxide at 200°C:

$$CuO(s) + H_2 = Cu(s) + H_2O(g)$$

Find $\Delta G°$, $\Delta H°$, and K for this reaction at 298°K. Calculate the equilibrium constant at 200°C, assuming that $\Delta H°$ is independent of temperature.

11. An electrochemical cell in which the spontaneous net reaction is

$$Zn(s) + 2H^+(aq) = Zn^{2+}(aq) + H_2(g)$$

develops 0.763 volt when all reagents are in their standard states at 298°K. (a) Calculate the standard free energy change for this reaction. (b) Calculate the standard free energy of $Zn^{2+}(aq)$ if the free energy of formation of $H^+(aq)$ is taken to be zero.

12. Once having defined the Gibbs free energy as $G = H - TS$, we found that $\Delta G < 0$ for any irreversible process at constant pressure and temperature. Another thermodynamic state function called the Helmholtz free energy can be defined by the expression $A = E - TS$. By following the methods used in the text to discuss G, show that for a system at constant volume and temperature, capable of doing only pressure-volume work, $\Delta A = 0$ for a reversible process and $\Delta A < 0$ for an irreversible process.

IV

Applications of the
Thermodynamic Principles

IN THIS CHAPTER we shall demonstrate the application of thermodynamics to some simple practical problems of interest to
chemists. Our purpose is to increase understanding of the thermodynamic principles and to show how thermodynamics helps us
to understand chemical phenomena.

4-1   PHASE EQUILIBRIA

Vapor Pressure

As our first application, we shall consider the temperature
dependence of the vapor pressure of a condensed material, such as
liquid water.

$$H_2O(l) = H_2O(g) \qquad \Delta H°_{298} = 10,510 \text{ cal/mole}$$

Generally, we refer to this, and any similar process, as a physical
change, but it is equally valid to think of it as the simplest possible
of chemical reactions. With this in mind we can write an equilibrium constant for the reaction:

$$K' = \frac{[H_2O(g)]}{[H_2O(l)]}$$

If we agree for the present to work with pure liquid water, then the concentration of liquid water is a constant and we can define a new equilibrium constant by the equation

$$K = K'[H_2O(l)] = [H_2O(g)]$$

This will be true as long as the water contains no dissolved solutes. Since the concentration of material in the gas phase is conveniently expressed in pressure units, we could better write the equilibrium constant as

$$K = P_{vap}$$

That is, the equilibrium constant for the reaction is the vapor pressure of water. A similar expression holds for the equilibrium between any liquid or solid and its vapor.

The relation between the vapor pressure of water and the standard free energy change $\Delta G°$ of evaporation is

$$\Delta G° = -RT \ln K = -RT \ln P_{vap}$$

We could evaluate the standard free energy change for the reaction from the standard entropies of liquid and gaseous water and the enthalpy of vaporization. However, in this case we shall reverse this procedure and obtain $\Delta G°$ from a direct measurement of the vapor pressure. At 298°K it is found that

$$P_{vap} = 23.76 \text{ mm}$$
$$= 0.03126 \text{ atm}$$
$$\Delta G° = -RT \ln P = (2.301)(1.987)(298) \log 0.03126$$
$$\Delta G° = 2065 \text{ cal}$$

This value of $\Delta G°$ could be used to calculate the standard free energy of formation of water vapor from that of liquid water. In turn, the free energy of formation of water vapor could be used in the calculation of other standard free energy changes. This is an example of how direct measurement of one equilibrium constant can yield information which can be used subsequently to help calculate many other equilibrium constants.

Now that we have defined the equilibrium constant for the evaporation process, we can use Eq. (3–26) to find how the vapor pressure of water depends on temperature. We have

$$\frac{d(\ln K)}{d(1/T)} = -\frac{\Delta H^\circ}{R} \qquad (3\text{–}26)$$

which becomes

$$\frac{d(\ln P_{vap})}{d(1/T)} = -\frac{\Delta H^\circ_{vap}}{R} \qquad (4\text{–}1)$$

If we assume ΔH_{vap} is independent of temperature and integrate, we find

$$\int d(\ln P_{vap}) = -\frac{\Delta H^\circ_{vap}}{R} \int d\left(\frac{1}{T}\right)$$

$$\ln P = -\frac{\Delta H^\circ_{vap}}{RT} + \text{const} \qquad (4\text{–}2)$$

$$\ln \frac{P_2}{P_1} = -\frac{\Delta H^\circ}{R}\left(\frac{1}{T_2} - \frac{1}{T_1}\right) \qquad (4\text{–}3)$$

Notice that Eq. (4–3) is also a direct result of the application of Eq. (3–29) to the vapor pressure problem. Since ΔH°_{vap} is always a positive number, Eq. (4–2) says that vapor pressure increases as temperature increases.

More precisely, a plot of the natural logarithm of vapor pressure as a function of reciprocal temperature should be a straight line of slope $-\Delta H^\circ_{vap}/R$. Figure 4–1 shows this type of plot for several liquids. While our prediction seems to be verified, careful examination of such plots shows that the lines are not exactly straight, but are curved slightly. This occurs because ΔH°_{vap} is not independent of temperature, as we have assumed. However, the temperature variation of ΔH°_{vap} is generally small enough that we are justified in using Eqs. (4–2) and (4–3) over small temperature ranges.

EXAMPLE 4–1

From the data given in the text, calculate the vapor pressure of water at 273°K. In Eq. (4–3) we have

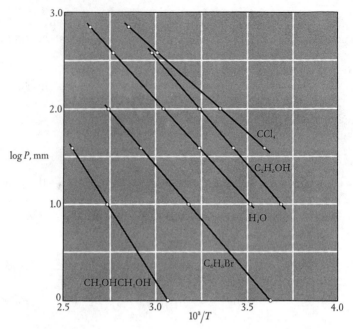

Figure 4–1 The logarithm of the vapor pressure of several liquids as a function of the reciprocal of absolute temperature.

$$T_2 = 273°K \qquad P_2 = ?$$
$$T_1 = 298°K \qquad P_1 = 23.8 \text{ mm}$$
$$\Delta H = 10{,}510 \text{ cal}$$
$$\ln \frac{P_2}{P_1} = 2.30 \log \frac{P_2}{P_1} = -\frac{10{,}510}{1.98}\left(\frac{1}{273} - \frac{1}{298}\right)$$
$$\log P_2 - \log 23.8 = -0.706$$
$$\log P_2 = 0.67$$
$$P_2 = 4.7 \text{ mm}$$

The measured value is 4.6 mm. A slight error is introduced in the calculation because $\Delta H_{\text{vap}}^{\circ}$ is not truly independent of temperature.

Equations (4–2) and (4–3) show that, if we are able to measure the vapor pressure at two or more temperatures, we can calculate ΔH_{vap} either directly from Eq. (4–3) or from a plot of $\ln P$ as a function of $1/T$. That is, we are able to obtain the enthalpy of vaporization without ever making a calorimetric measurement.

In order to provide a clearer understanding of its origins, we shall derive Eq. (4–1) again by using a somewhat different procedure. Consider a liquid and its vapor in equilibrium in a closed vessel. The condition for equilibrium between the two phases is that the free energy of a mole of liquid be the same as that of a mole of vapor:

$$\bar{G}_l = \bar{G}_g \qquad (4\text{--}4)$$

If we now make infinitesimal changes in the temperature and pressure, the free energy of the liquid and vapor change by amounts dG_l and dG_g, respectively. If the system is to remain at equilibrium, it must be that

$$d\bar{G}_l = d\bar{G}_g$$

Now we use Eq. (3–12) to express the changes in free energy in terms of pressure and temperature:

$$d\bar{G}_l = \bar{V}_l \, dP - \bar{S}_l \, dT = d\bar{G}_g = \bar{V}_g \, dP - \bar{S}_g \, dT$$

$$(\bar{V}_l - \bar{V}_g) \, dP = (\bar{S}_l - \bar{S}_g) \, dT$$

where $\bar{V}$ and $\bar{S}$ are the volume and entropy per mole of material.

$$\frac{dP}{dT} = \frac{\bar{S}_g - \bar{S}_l}{\bar{V}_g - \bar{V}_l} = \frac{\Delta S}{\Delta V}$$

Since the two phases are at equilibrium, $\Delta G = 0$ and $T \, \Delta S = \Delta H$. Therefore,

$$\frac{dP}{dT} = \frac{\Delta H}{T \, \Delta V} \qquad (4\text{--}5)$$

This important relation is called the **Clapeyron equation.** Although we have derived it with the special case of liquid-vapor equilibrium in mind, we shall see that it can be applied to the equilibrium between any two phases. To see how the Clapeyron equation leads to Eq. (4–1), we note that the molar volume of a gas is always considerably larger than the molar volume of a liquid. Therefore, we can justifiably make the approximation that

$$\Delta \bar{V} = \bar{V}_g - \bar{V}_l \cong \bar{V}_g = \frac{RT}{P}$$

Using this approximation, we find that Eq. (4–5) becomes

$$\frac{dP}{dT} = \frac{P \, \Delta H^\circ_{vap}}{RT^2}$$

$$\frac{1}{P}\frac{dP}{dT} = \frac{d(\ln P)}{dT} = \frac{\Delta H^\circ_{vap}}{RT^2}$$

or

$$\frac{d(\ln P)}{d(1/T)} = -\frac{\Delta H^\circ_{vap}}{R}$$

This is the same as Eq. (4–1), found earlier. This latter method of derivation has the advantage that it emphasizes Eq. (4–4), the fundamental condition for phase equilibrium: the molar free energy of a substance must be the same in both phases.

As we have already mentioned, Eq. (4–2) also applies to the equilibrium of a solid and its vapor. The evaporation of a solid is called sublimation, and the ΔH for this process is known as the enthalpy of sublimation. Equation (4–2) then shows that a plot of the natural logarithm of the vapor pressure of a solid as a function of reciprocal temperature is a straight line whose slope is $-\Delta H^\circ_{sub}/R$. The enthalpies of sublimation, vaporization, and fusion are related by the equation

$$\Delta H^\circ_{sub} = \Delta H^\circ_{fus} + \Delta H^\circ_{vap} \tag{4–6}$$

Therefore, if we plot the logarithm of vapor pressure as a function of reciprocal temperature over a temperature range which includes both the solid and liquid phases, we should find two straight lines whose slopes should be $\Delta H^\circ_{sub}/R$ and $\Delta H^\circ_{vap}/R$. An example of this is shown in Fig. 4–2. The difference in the slopes should be, by Eq. (4–6), $\Delta H^\circ_{fus}/R$. Thus, by measuring vapor pressures over a

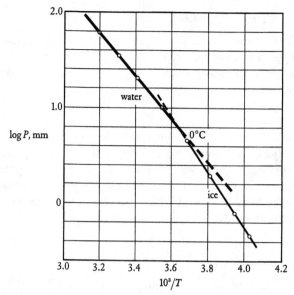

Figure 4-2 **The logarithm of the vapor pressure of ice and water as a function of reciprocal temperature. The difference in slopes represents the difference between ΔH_{vap} and ΔH_{sub}.**

wide range of temperatures, it is possible to obtain ΔH°_{sub}, ΔH°_{vap}, and ΔH°_{fus}.

The two straight lines in Fig. 4-2 intersect at the temperature at which solid and liquid have the same vapor pressure. Since at this temperature solid and liquid are in equilibrium with the same pressure of the vapor, the solid and liquid must be in equilibrium with each other. There is only one value of the temperature and pressure at which the three pure phases are simultaneously in equilibrium, and this temperature is called the triple point. The triple point differs only slightly from what is called the normal melting point—the temperature at which solid, liquid, and vapor are in equilibrium in the *presence of air at one atmosphere pressure*. For example, the melting point of ice is 0°C, and its triple point is 0.0099°C.

Condensed Phase Equilibria

The argument used to derive the Clapeyron relation, Eq. (4–5), applies not only to vaporization processes but to all phase changes. For instance, when a solid is in equilibrium with its liquid, we have

$$\bar{G}_s = \bar{G}_l$$

If the solid-liquid system completely fills a closed vessel which has an adjustable volume, we can study the effect of pressure on the temperature at which solid and liquid are in equilibrium. Equation (4–5) says that an infinitesimal pressure change dP changes the temperature at which liquid and solid are in equilibrium by an amount

$$dT = \frac{T \, \Delta V}{\Delta H} \, dP$$

where ΔH and ΔV are the enthalpy and volume changes for the process

$$\text{solid } (T) = \text{liquid } (T)$$

The enthalpy of the melting process is always positive, but ΔV may be positive or negative. If the molar volume of the liquid is less than that of the solid, as is the case for the ice-water transformation, then ΔV is negative; and dT is negative if dP is positive. In other words, the melting point of ice and of other such solids goes down as the applied pressure increases. The more common situation is to have $\bar{V}_l > \bar{V}_s$; then the melting temperature of the solid increases as pressure increases.

EXAMPLE 4–2

Calculate the change in the melting point of ice which is brought about by application of 100 atm pressure. The volume change upon melting is -1.63×10^{-3} liter/mole, and the enthalpy of fusion is 1440 cal/mole.

The Clapeyron relation, Eq. (4–5), may be written as

$$dT = T \frac{\Delta V}{\Delta H} dP$$

It is reasonable to assume that the ΔV and ΔH of melting are independent of pressure and write

$$\int dT = T \frac{\Delta V}{\Delta H} \int dP$$

$$\Delta T = T \frac{\Delta V}{\Delta H} \Delta P$$

Before substituting the given information, we must take care that the units to be used are homogeneous. A little reflection shows that $\Delta V \, \Delta P$ and ΔH must both be expressed in the same units. Since 1 liter-atm = 24.4 cal, we have

$$\Delta T = (273°) \frac{(-1.63 \times 10^{-3} \text{ liter})(100 \text{ atm})(24.4 \text{ cal/liter-atm})}{1440 \text{ cal}}$$

$$\Delta T = -0.75°$$

At 100 atm pressure, ice melts at $-0.75°$C.

4–2 PROPERTIES OF IDEAL SOLUTIONS

Our application of thermodynamics to phase equilibria has been exclusively concerned with the behavior of pure substances. However, it is not at all difficult to extend the arguments to the properties of solutions. We shall consider only dilute solutions of a nonvolatile solute such as sugar in a volatile solvent like water. If these solutions are sufficiently dilute, the vapor pressure of the solvent follows Raoult's law:

$$P = x_1 P° \tag{4–7}$$

In this expression P is the vapor pressure of the solvent in solution, x_1 is its mole fraction, and $P°$ is the vapor pressure of the pure solvent. As shown in Fig. 4–3, the vapor pressure of the solution is lower than that of the pure solvent at all temperatures. As a result of this vapor pressure lowering, the solution boils at a higher temperature and freezes at a lower temperature than the pure solvent. We now wish to relate the change in boiling and freezing temperatures to the concentration of the solution and properties of the solvent.

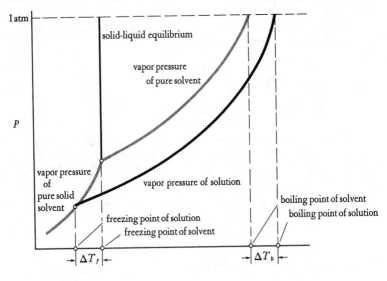

Figure 4–3 A comparison of the vapor pressure of a pure solvent and its solution showing the origin of freezing point depression and boiling point elevation.

Boiling Point Elevation

Let us begin with the concentration dependence of the boiling point of an ideal solution. We are studying the reaction

$$\text{liquid(conc } x_1) = \text{vapor(1 atm)}$$

The equilibrium constant for this reaction is

$$K' = \frac{[\text{vapor}]}{[\text{liquid}]} = \frac{[1 \text{ atm}]}{[x_1]}$$

We are interested only in the variation of the *normal boiling point* of the solution with temperature. Therefore, the concentration of the vapor is *always* 1 atm, and we can define a new equilibrium constant by

$$K = \frac{K'}{[1 \text{ atm}]} = \frac{1}{[x_1]} \qquad (4\text{-}8)$$

Our general equation for the temperature variation of an equilibrium constant is

$$\frac{d(\ln K)}{dT} = \frac{\Delta H^\circ}{RT^2} \qquad (3\text{-}27)$$

Applying this to Eq. (4–8) gives us

$$\frac{d[\ln (1/x_1)]}{dT} = -\frac{d(\ln x_1)}{dT} = \frac{\Delta H^\circ_{\text{vap}}}{RT^2}$$

$$d(\ln x_1) = -\frac{\Delta H^\circ_{\text{vap}}}{RT^2} dT \qquad (4\text{-}9)$$

Equation (4–9) says that the boiling point of a solution changes by an infinitesimal amount dT when the concentration of the solution is changed by an amount represented by $d(\ln x_1)$. In order to find the solvent concentration x_1, which corresponds to a boiling temperature T, we integrate Eq. (4–9), using as lower limits the conditions which correspond to *pure solvent:* $x_1 = 1$ ($\ln x_1 = 0$) and a boiling temperature T_b. Thus

$$\int_{\ln x_1 = 0}^{\ln x_1} d(\ln x_1) = -\frac{\Delta H^\circ_{\text{vap}}}{R} \int_{T_b}^{T} \frac{dT}{T^2}$$

$$\ln x_1 = \frac{\Delta H^\circ_{\text{vap}}}{R} \left(\frac{1}{T} - \frac{1}{T_b} \right)$$

$$\ln x_1 = \frac{\Delta H^\circ_{\text{vap}}}{R} \frac{(T_b - T)}{TT_b}$$

Notice that these equations can be obtained simply by substituting the values $K_1 = 1$, $T_1 = T_b$, $K_2 = 1/x$, and $T_2 = T$ into Eq. (3–29).

Let us now define the boiling point elevation ΔT by the expression

$$\Delta T = T - T_b$$

and realize that when the solution is dilute, ΔT is small, and $T \cong T_b$. Therefore, setting TT_b equal to T_b^2, we write

$$\ln x_1 = -\frac{\Delta H_{vap}^{\circ}}{RT_b^2} \Delta T \qquad\qquad (4\text{--}10)$$

This expression relates the elevation of the boiling point ΔT to the concentration of the solvent. We can simplify it somewhat by recalling that for a two-component mixture

$$x_1 = 1 - x_2$$

where x_2 is the mole fraction of the solute. Consider also that

$$\ln x_1 = \ln(1 - x_2) = -x_2 - \tfrac{1}{2}x_2^2 + \cdots$$

The first term of this series expansion is a satisfactory approximation of $\ln x_1$ if x_2 is small, as it is in dilute solutions. After a little rearrangement, Eq. (4–10) becomes

$$\Delta T = \frac{RT_b^2}{\Delta H_{vap}^{\circ}} x_2 \qquad\qquad (4\text{--}11)$$

It is more common to express the concentration in units of molality m, which is the number of moles of solute in 1000 g of solvent. To change x_2 to molality, we take advantage of the fact that the solution is dilute and write

$$x_2 = \frac{n_2}{n_1 + n_2} \cong \frac{n_2}{n_1} \qquad\qquad (\text{if } n_2 \ll n_1)$$

The number of moles of solvent n_1 is equal to w_1/M_1, where w_1 and M_1 are the weight and molecular weight of solvent, respectively. Also, n_2/w_1 is the number of moles of solute per gram of solvent, so the molality is

$$\frac{n_2}{w_1} 1000 = m$$

and so

$$x_2 \cong \frac{n_2}{n_1} = \frac{n_2}{w_1} M_1 = \frac{mM_1}{1000}$$

Therefore, Eq. (4–11) becomes

Table 4-1
Boiling Point Elevation Constants

	T_b, °K	$\Delta H_{vap.}$ kcal/mole	$K_b(calc)$	$K_b(exp)$
Water	373.1	9.72	0.51	0.51
Benzene	353.3	7.35	2.62	2.53
Acetone	329.1	7.34	1.70	1.71
Chloroform	334.3	7.02	3.67	3.63

$$\Delta T = \left[\frac{RT_b^2 M_1}{1000\ \Delta H_{vap}^\circ} \right] m \qquad (4\text{-}12)$$

$$\Delta T = K_b m$$

The constant K_b called the molal boiling point elevation constant, depends, as Eq. (4-12) shows, only on properties of the solvent. Table 4-1 gives K_b for several solvents. The experimental values of K_b are obtained by direct measurement of ΔT for solutions of known molality. The calculated values are found from the boiling temperature and enthalpy of vaporization by use of Eq. (4-12).

EXAMPLE 4-3

Calculate the molal boiling point elevation constant for water, which has $T_b = 373°K$ and $\Delta H_{vap}^\circ = 9720$ cal.

Since ΔH_{vap}° is given in calories, R in Eq. (4-12) must also be expressed in calories.

$$K_b = \frac{(1.98)(373)^2(18.0)}{(1000)(9720)} = 0.513 \text{ deg/molal}$$

EXAMPLE 4-4

Calculate the boiling point of a solution containing 30.0 g of sugar (molecular weight 342) in 100 g of water.

The molality is

$$m = \frac{30.0}{342} \times \frac{1000}{100} = 0.878$$

Therefore,

$$\Delta T = K_b m = (0.513)(0.878) = 0.451$$

The boiling point of the solution is $100.45°C$.

In order to make the origin of the boiling point elevation more clear, we offer another derivation of Eq. (4–11). We start with Eq. (4–1), which relates changes of vapor pressure and temperature:

$$d(\ln P) = -\frac{\Delta H°_{vap}}{R} d\left(\frac{1}{T}\right) \qquad (4\text{–}1)$$

But $d(\ln P) = dP/P$ and $d(1/T) = -dT/T^2$, so

$$\frac{dP}{P} = \frac{\Delta H°_{vap}}{RT^2} dT \qquad (4\text{–}13)$$

If we make a small change ΔP in the vapor pressure, the boiling temperature will change by a small amount ΔT. The relation between these changes can be found by integrating Eq. (4–13) over a small range in which P and T may be considered constant and kept outside the integral sign.

$$\frac{1}{P}\int dP = \frac{\Delta H°_{vap}}{RT^2} \int dT$$

$$\frac{1}{P}\Delta P = \frac{\Delta H°_{vap}}{RT^2} \Delta T \qquad (4\text{–}14)$$

Since we are treating dilute solutions, we can obtain an expression for $\Delta P/P$ by using Raoult's law:

$$P_1 = x_1 P_1°$$
$$\Delta P = P_1° - P_1 = P_1° - x_1 P_1° = (1 - x_1)P_1°$$
$$\Delta P = x_2 P_1°$$

Therefore the left side of Eq. (4–14) becomes $\Delta P/P = x_2$ and

$$\frac{\Delta P}{P} = x_2 = \frac{\Delta H°_{vap}}{RT^2} \Delta T$$

$$\Delta T = \frac{RT^2}{\Delta H°_{vap}} x_2$$

This is the same as Eq. (4–11). This latter derivation emphasizes the connection between boiling point elevation and vapor pressure lowering, while the first derivation has the advantage of closely following the procedures we used to find Eq. (3–29), the general temperature dependence of an equilibrium constant.

Freezing Point Depression

Figure 4–3 shows that the lowering of the vapor pressure of a solution is closely connected to the depression of the freezing point. The easiest way to derive the relationship between freezing point, concentration, and properties of the solvent is to realize we are studying the reaction

$$\text{solid(pure)} = \text{liquid(conc } x_1)$$

Since the concentration of the pure solid is a constant, we can formally write the equilibrium constant for the reaction as

$$K = x_1$$

Our general expression for the temperature variation of the equilibrium constant is

$$\frac{d(\ln K)}{dT} = \frac{\Delta H^\circ}{RT^2}$$

which, when applied to the solid-solution equilibrium, becomes

$$\frac{d(\ln x_1)}{dT} = \frac{\Delta H^\circ_{\text{fus}}}{RT^2} \qquad (4\text{--}15)$$

where $\Delta H^\circ_{\text{fus}}$ is the enthalpy of fusion. The derivation proceeds in a manner exactly analogous to that used for the boiling point elevation. We integrate Eq. (4–15) from T_f, the temperature at which the pure solid is in equilibrium with its pure liquid, to T, the temperature at which the pure solid is in equilibrium with a solution of concentration x_1. This gives us

$$\int_{x_1=1}^{x_1} d(\ln x_1) = \frac{\Delta H^\circ_{\text{fus}}}{R} \int_{T_f}^{T} \frac{dT}{T^2}$$

$$\ln x_1 = -\frac{\Delta H^\circ_{\text{fus}}}{R}\left(\frac{1}{T} - \frac{1}{T_f}\right)$$

$$\ln x_1 = -\frac{\Delta H^\circ_{\text{fus}}}{R}\frac{(T_f - T)}{TT_f}$$

If we define $\Delta T = T_f - T$ as the freezing point lowering and make the approximation that $TT_f = T_f^2$, then

$$\ln x_1 = -\frac{\Delta H^\circ_{\text{fus}}}{RT_f^2}\Delta T$$

Expansion of the logarithm as before gives

$$\Delta T = \frac{RT_f^2}{\Delta H^\circ_{\text{fus}}}x_2$$

Conversion to molality leads us to the equation

$$\Delta T = \left(\frac{RT_f^2 M_1}{1000\,\Delta H^\circ_{\text{fus}}}\right)m \qquad (4\text{--}16)$$

$$\Delta T = K_f m \qquad (4\text{--}17)$$

Here T_f is the melting temperature of the pure solid, M_1 is its molecular weight, and $\Delta H^\circ_{\text{fus}}$ is its enthalpy of fusion. Therefore, the freezing point depression constant K_f depends only on the properties of the solid, which, since it is in excess, acts as the solvent in the solution.

Freezing point depression offers a convenient way to estimate the molecular weight of new compounds. If a weighed amount of an unknown compound is dissolved in a material whose K_f is known, the measured freezing point depression gives the molality of the unknown solution, which then allows calculation of M_2, the molecular weight of the unknown. Solids which have large values of K_f are particularly useful in these measurements, since then only a small amount of unknown is required to produce an easily measurable ΔT. Table 4–2 contains the K_f for some common solvents. A comparison of the experimentally observed K_f with that calculated from ΔH_{fus} and T_f demonstrates the validity of our derivation.

Table 4-2
Freezing-Point-Depression Constants

	T_f, °K	$\Delta H_{fus,}$ kcal/mole	$K_f(calc)$	$K_f(exp)$
Water	273.1	1.44	1.85	1.86
Benzene	278.5	2.35	5.09	5.12
Napthalene	353.1	4.55	6.9	6.8
Camphor	451.1	1.38	47.5	48.5

EXAMPLE 4-5

Calculate the freezing point of a solution of 0.060 g of NaCl in 100 g of water. The freezing point depression constant for water is 1.85 deg/molal.

The molality of NaCl is

$$\frac{0.060}{58.5} \times \frac{1000}{100} = 0.0103$$

Since NaCl is completely dissociated in solution, the total molality is $2 \times 0.0103 = 0.021$. Therefore,

$$\Delta T = (1.85)(0.021) = 0.039°$$

Thus the freezing point of the solution is $-0.039°C$.

Solubility of Solids

The temperature dependence of the solubility of a solid is closely related to the problem of freezing point depression. The reaction is

pure solid solute = solute in solution (conc x_2)

Since the concentration of the pure solid solute is a constant, the equilibrium constant for the reaction is simply

$$K = x_2$$

The temperature dependence of the equilibrium constant is once again

$$\frac{d(\ln K)}{d(1/T)} = -\frac{\Delta H°}{R}$$

To proceed, we must know what $\Delta H°$ for the reaction is. The answer appears easily if we imagine the solution process occurring in two steps:

solid solute → liquid solute → solute in solution

The enthalpy change for the first step is clearly $\Delta H°_{fus}$, the enthalpy of fusion of the pure solute. *If the solution is ideal,* the enthalpy of the second step is zero. This statement is really the definition of an ideal solution; if two liquids are mixed and no heat is evolved or absorbed, the resulting solution is ideal. Therefore, restricting ourselves to ideal solutions, we can write

$$\frac{d(\ln K)}{d(1/T)} = \frac{d(\ln x_2)}{d(1/T)} = -\frac{\Delta H°_{fus}}{R}$$

$$\int d(\ln x_2) = -\frac{\Delta H°_{fus}}{R} \int d\left(\frac{1}{T}\right)$$

$$\ln x_2 = -\frac{\Delta H°_{fus}}{RT} + \text{const} \qquad\qquad \textbf{(4-18)}$$

This equation relates the equilibrium concentration of solute x_2 to the temperature of the solution. Since ΔH_{fus} is always positive, Eq. (4-18) predicts that the solubility should increase as temperature increases. In fact, plotting the natural logarithm of the solubility as a function of reciprocal temperature should yield a straight line of slope $-\Delta H°_{fus}/R$. Figure 4-4 shows this is realized for solutions of naphthalene in benzene, toluene, and nitrobenzene. For solutions which are not ideal, the temperature dependence of solubility is more complicated.

Comparison of the treatment of temperature dependence of solubility with that of freezing point depression shows that the two are very similar. This is not an accident, since both processes involve conversion of a pure solid into a solution. In the case of freezing point depression, the solid "melts" and enters a solution as the *solvent,* since it is in excess. In the problem of temperature dependence of solubility, the solid enters a solution in which it is in

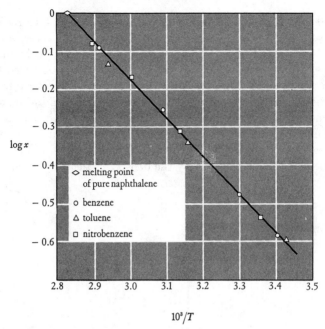

$$10^3/T$$

Figure 4-4 The logarithm of the solubility of naphthalene, expressed in mole fraction units, as a function of the reciprocal of absolute temperature.

minor concentration, and thus it is called the solute. There is no physical difference between the two phenomena, and our mathematical treatments of them are up to a point identical, except that we represent the concentration of the *solid* in solution by x_1 when it is in excess and by x_2 when it is not.

Osmotic Pressure

As our last example of the equilibria associated with solutions we turn to the problem of osmotic pressure. The phenomenon of osmotic pressure involves a semipermeable membrane, that is, some film which has pores large enough to allow the passage of small solvent molecules but small enough to prevent high molecular weight solute molecules from passing through. When a solution is separated from its pure solvent by a semipermeable membrane, as in the apparatus shown in Fig. 4–5, it is observed that some of the pure solvent passes through the membrane into the solution. The flow stops and the system reaches equilibrium after the meniscus has risen to a height which depends on the concentration of the solution. Under these conditions the solution is under a greater hydrostatic pressure than the pure solvent. The height of the meniscus, multiplied by the density of the solution and the

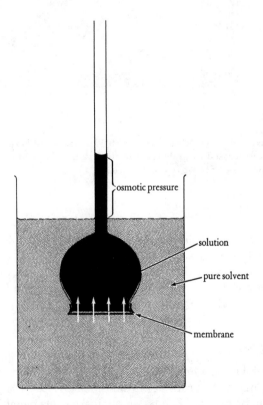

osmotic pressure

solution

pure solvent

membrane

Figure 4–5 A simple apparatus showing the osmotic pressure phenomenon.

acceleration of gravity, gives the excess pressure on the solution, and this is the osmotic pressure.

In order to understand this phenomenon, it is first necessary to realize that the only role of the semipermeable membrane is to allow the solvent in the solution to come to equilibrium with the pure solvent. This equilibrum will be reached only when the molar free energy of the solvent *in solution* is the same as that of the pure solvent. Our previous experience has shown that the free energy of the solvent in solution is less than that of the pure solvent. That is, the molar free energy of the solvent of mole fraction x_1 is

$$\bar{G} = \bar{G}° + RT \ln x_1$$

where $\bar{G}°$ is the free energy of the pure solvent.

Since the free energy of the solvent in solution is less than that of the pure solvent, the pure solvent will have a tendency to flow through the semipermeable membrane into the solution. This flow must continue until some other factor raises the free energy of the solvent in the solution back to the value $\bar{G}°$. This other factor is the osmotic pressure on the solution. According to Eq. (3–12), the effect of pressure on free energy is given by

$$dG = V \, dP$$

for a process at constant temperature; so raising the pressure increases free energy. Since the osmotic pressure π is the pressure on the solution in excess of that exerted on the pure solvent, the increase of the molar free energy resulting from osmotic pressure is

$$\Delta\bar{G} = \bar{V} \int_0^\pi dP = \pi \bar{V}$$

Here we have assumed that the molar volume $\bar{V}$ of the solvent in solution is independent of pressure.

Now we can express the combined effects of dilution and pressure on the free energy of the solvent by writing

$$\bar{G} = \bar{G}° + RT \ln x_1 + \pi \bar{V}$$

When the solvent in solution is in equilibrium with the pure solvent, $\bar{G} = \bar{G}°$ and

$$\pi V = -RT \ln x_1$$

Substituting $x_1 = 1 - x_2$ and expanding the logarithm as before gives

$$\pi \overline{V} = RT\, x_2$$

If the solution is dilute, $x_2 \cong n_2/n_1$ and $\overline{V} = V/n_1$, where V is the total volume of the solution and n_1 and n_2 are the number of moles of solvent and solute, respectively. Therefore

$$\pi \frac{V}{n_1} = RT \frac{n_2}{n_1}$$

$$\pi = \frac{RT}{V} n_2 \qquad\qquad (4\text{--}19)$$

The final expression for the osmotic pressure in terms of the number of moles n_2 of solute and the total volume V of the solution has a remarkable resemblance to the ideal-gas equation of state.

The osmotic pressure corresponding to a solute concentration of 1 mole/liter would be

$$\pi = 0.082 \frac{\text{liter-atm}}{\text{mole-deg}} \times 298 \text{ deg} \times \frac{1 \text{ mole}}{1 \text{ liter}} = 22.4 \text{ atm}$$

Since it is possible to detect and measure precisely pressures less than 10^{-3} atm, the osmotic pressure due to 10^{-4} moles of solute per liter of solvent is easily detectable. The great sensitivity of osmotic pressure is used to advantage in the determination of the molecular weight of high polymers, particularly biological molecules. These substances tend to be very insoluble; however, it is often possible to measure the osmotic pressure of their very dilute solutions and, by knowing the weight of material dissolved, calculate their molecular weight.

EXAMPLE 4-6

A sample of polyisobutylene weighing 0.520 g was dissolved in exactly 100 ml of benzene and placed in an osmometer. The solution came to equilibrium with the pure solvent when the osmotic pressure on the solution was 4.12×10^{-3} atm at a temperature of 298°K. What is the molecular weight of the polyisobutylene?

Inserting the data into Eq. (4-19), we have

$$\pi = \frac{RT}{V} n_2$$

$$4.12 \times 10^{-3} = \frac{(0.0820)(298)}{0.100} \frac{0.520}{M}$$

$$M = 31,000$$

4-3 CHEMICAL APPLICATIONS

In order to demonstrate how thermodynamic analysis can aid the understanding of chemical behavior, we turn to several examples of the thermodynamic treatment of chemical equilibria.

Temperature Dependence of Equilibria

As our first application we choose a reaction that is important in the petroleum industry, the isomerization of normal butane to isobutane:

$$
\underset{\text{\textit{n}-butane}}{CH_3CH_2CH_2CH_3} \rightarrow \underset{\text{isobutane}}{CH_3 - \overset{\overset{\displaystyle CH_3}{|}}{\underset{\underset{\displaystyle H}{|}}{C}} - CH_3}
$$

Isobutane is one of the important starting materials in the synthesis of the hydrocarbons which are components of high-performance gasoline. Therefore, we seek conditions under which we can convert the naturally occurring *n*-butane to isobutane with as high yield as possible. To find these conditions, we must evaluate the equilibrium constant of the reaction and its temperature dependence.

From Tables 2-2 and 3-1 we obtain:

n-butane	isobutane
$S° = 74.10$ eu	$S° = 70.42$ eu
$\Delta H_f° = -29,810$ cal	$\Delta H_f° = -31,450$ cal

Therefore, $\Delta H° = -1640$ and $\Delta S° = -3.68$ for the reaction as written. These data show that isobutane is energetically more stable than n-butane, and this factor should favor the conversion we desire. On the other hand, the entropy of isobutane is less than that of n-butane, and this factor will tend to diminish the equilibrium constant of the reaction. We can assess the relative importance of the two effects by calculating the equilibrium constant using Eq. (3–24).

$$\ln K = 2.303 \log K = \frac{\Delta S°}{R} - \frac{\Delta H°}{RT} \qquad (3\text{--}24)$$

$$K = 10^{\Delta S°/2.3R} \times 10^{-\Delta H°/2.3RT} \qquad (4\text{--}20)$$

Insertion of the values of $\Delta H°$ and $\Delta S°$ obtained above gives us at 298°K

$$K = 10^{-0.8} \times 10^{1.2} = 10^{0.4} = 2.5$$

As we have noted, the entropy decrease tends to diminish the equilibrium constant, while the enthalpy provides the driving force for the formation of isobutane. These opposing factors are nearly equal at 298°K, and it is clear from Eq. (4–20) that the equilibrium constant will become smaller as temperature increases. Therefore, if this reaction is to be carried out successfully, it must be run at as low a temperature as is possible. However, in the absence of catalysts the reaction is immeasurably slow at room temperature.

The situation might seem hopeless, since raising the temperature to increase the reaction rate would decrease the equilibrium constant, and thus the maximum product yield would approach zero. The problem is solved by the use of sulfuric and hydrofluoric acid catalysts that permit the reaction to proceed rapidly to equilibrium at room temperature. It was thermodynamic analysis such as we have just made which helped to avoid fruitless attempts to run this reaction at high temperature and directed effort toward development of a low-temperature catalytic process.

Before leaving this example, let us try to rationalize the differences in enthalpy and entropy which exist between these two very similar molecules. The slightly lower ΔH_f° of isobutane must mean that the bonds in this molecule are slightly stronger than those in *n*-butane. Yet in each compound there are three C—C bonds and ten C—H bonds. The difference in enthalpy between *n*-butane and isobutane must then be a consequence of the fact which we pointed out in the section devoted to bond energy: the energy of a particular type of bond is not strictly constant, but varies slightly depending on the nature of its surroundings in a molecule. We must conclude, then, that the energies of the bonds involved in CH_2 groups such as are in *n*-butane are slightly different from those involved in CH_3 or CH groups in isobutane. These differences are so small that it is difficult to predict them from molecular theory, but we can expect to encounter these same differences in other hydrocarbon molecules.

The problem of understanding the difference in entropy between isobutane and *n*-butane is more tractable. We have said that entropy is a measure of the number of microscopic states available to a system, or that it is measure of the molecular disorder in a system. The reason that the entropy of *n*-butane is greater than that of isobutane is that the atoms in *n*-butane have available to them a greater number of geometrical arrangements than those in isobutane have. This extra freedom in *n*-butane occurs because it is possible for this molecule to coil back on itself simply by rotating one half of the molecule about the central C—C bond. This is shown in Fig. 4–6. All positions between the extremes that appear in the figure are also possible. Isobutane does

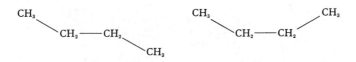

Figure 4–6 Two configurations of the *n*-butane molecule. Since rotation about the central bond is possible, the carbon atoms have some freedom to move relative to each other.

not have this extra flexibility, since it has no chain more than three carbon atoms long. Figure 4–7 shows that there is no possibility for coiling this molecule back upon itself, and the carbon atoms are held in well-defined positions relative to each other. Thus isobutane does not have the internal freedom and extra entropy associated with *n*-butane.

This interpretation of entropy in terms of the variety of internal configurations can be applied to other molecules. For example, the entropy of cyclohexane is 71.28 eu, whereas the entropy of its isomer 1-hexene is 92.25 eu. As shown in Fig. 4–8, the 1-hexene isomer has an opportunity to assume a variety of coiled or extended configurations, whereas in the cyclohexane molecule binding the atoms in a ring reduces their freedom considerably.

As a second example of the application of thermodynamics to chemical equilibrium we choose the problem of converting 1-butene into 1,3-butadiene by elimination of hydrogen.

$$CH_2{=}CH{-}CH_2CH_3 \longrightarrow CH_2{=}CH{-}CH{=}CH_2 + H_2$$

1-butene   **1,3-butadiene**

This process is one step in the conversion of petroleum products to synthetic rubber. To determine the conditions under which the

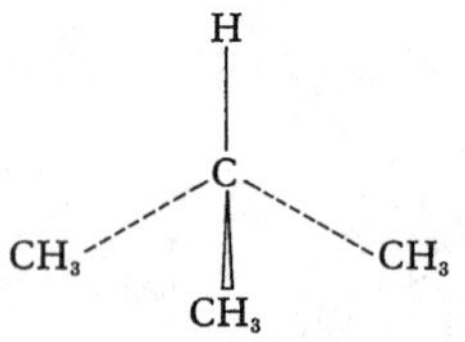

Figure 4–7 The configuration of the isobutane molecule. The carbon atoms are fixed in space relative to one another.

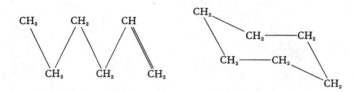

Figure 4–8 Configurations of the isomers 1-hexene and cyclohexane. The ends of the 1-hexene molecule can move relative to each other, while the motion in cyclohexane is considerably restricted.

yield of butadiene will be appreciable, we again consult Tables 2–2 and 3–1 and find that, at 298°K:

1-butene	**1,3-butadiene**	**hydrogen**
$S° = 73.48$ eu	$S° = 66.62$ eu	$S° = 31.21$ eu
$\Delta H° = 280$ cal	$\Delta H°_f = 26{,}750$ cal	$\Delta H° = 0$

For the reaction $\Delta S° = 24.3$ and $\Delta H° = 26{,}470$; therefore, the entropy change favors the products, but the reactant is energetically more stable. At 298°K we find the equilibrium constant is

$$K = 10^{\Delta S°/2.3R} \times 10^{-\Delta H°/2.3RT}$$
$$K = 10^{5.3} \times 10^{-19.4} = 1.2 \times 10^{-15}$$

That is, if the pressure of 1-butene were 1 atm, the equilibrium pressure of butadiene and hydrogen would be only 3×10^{-7} atm, a discouragingly small number. However, the reaction is endothermic, and therefore an increase in temperature should increase the equilibrium constant. We can obtain some idea of the value of the equilibrium constant at elevated temperatures, say 1000°K, by applying Eq. (3–29), assuming $\Delta H°$ is independent of temperature:

$$\ln \frac{K_2}{K_1} = -\frac{\Delta H^\circ}{R}\left(\frac{1}{T_2} - \frac{1}{T_1}\right) \qquad (3\text{--}29)$$

$$2.3 \log K_2 - 2.3 \log (1.2 \times 10^{-15}) = -\frac{26,500}{1.98}\left(\frac{1}{1000} - \frac{1}{298}\right)$$

$$\log K_2 = -0.45$$

$$K_2 = 0.35$$

Considering the fact that we assumed that ΔH° was independent of temperature over such a wide temperature range, our answer compares very favorably with a more carefully obtained value of unity. Even though our calculation was crude, it did give us a reasonable estimate of the conditions necessary to effect appreciable conversion of reactant to product.

Finally, let us interpret the values of ΔH° and ΔS° by using our knowledge of molecular structure. In the reaction we break two C—H bonds and one C—C bond and we form one H—H bond and one C=C bond. We expect the ΔH° of the reaction to be the difference between the energies of the bonds broken and made. Using Table 2–3, we obtain

$$\Delta H^\circ = -\Sigma\,\epsilon(\text{bonds formed}) + \Sigma\,\epsilon(\text{bonds broken})$$
$$= -\epsilon(\text{C}{=}\text{C}) - \epsilon(\text{H—H}) + 2\epsilon(\text{C—H}) + \epsilon(\text{C—C})$$
$$= -145 - 104 + 196 + 80$$
$$\Delta H^\circ = 27 \text{ kcal}$$

This result compares very favorably with the experimentally observed value of 26.5 kcal. Actually, this particular answer is better than we might expect, since 1,3-butadiene is a molecule which, because of the location of two double bonds separated by one single bond, has double bond energies which are slightly atypical. Fortunately, this effect is small, so in this particular case our crude estimate of ΔH° based on average bond energies is satisfactory.

Now we attack the easier problem of interpreting the entropy change of the reaction. The large positive value of 24.3 eu means that the chaos, disorder, or freedom of the atoms in the products is larger than in the reactants. We can readily understand this, since in the reaction two hydrogen atoms originally localized on a butene molecule have been set free as a hydrogen molecule. That is, hydrogen atoms attached to a carbon chain are in a state of higher order than when they are free to move independently of the carbon chain. With this model system in mind, we can correctly expect all gaseous dissociation reactions to have positive entropy changes. Naturally, reactions in which molecules are completely dissociated to their constituent gaseous atoms have the largest entropy increase, and it is only the accompanying increase in enthalpy which prevents molecules from completely breaking apart.

Until now we have been almost exclusively occupied with the application of thermodynamics to gaseous reactions. We have been so in order to take advantage of the simplicity of gaseous systems, not because of any limitation of thermodynamic principles. Now that we have a firmer idea of the nature of entropy and enthalpy, we are in a position to see what thermodynamic analysis of solution reactions can teach us.

Weak Acid Ionization

In elementary textbooks one can find attempts to predict or explain the magnitudes of the ionization constants of weak acids. In most cases the arguments are based exclusively on a qualitative assessment of the amount of energy necessary to remove a proton from the parent acid to form a hydrated proton and the negative ion. For example, the ionization of phosphoric acid is often written as

$$H_3PO_4 = H^+ + H_2PO_4^-$$

If we interpret this equation literally, the existence of separated positive and negative charges on the right side of the equation suggests that a large amount of energy should be required to effect the dissociation and that an increase in entropy resulting from the greater freedom of the proton is responsible for producing any dissociation whatsoever. More careful examination of the process shows that the opposite is true. From Table 4–3 we obtain

$H_3PO_4(aq)$ $H_2PO_4^-(aq)$
$S^\circ = 42.1$ eu $S^\circ = 21.3$ eu
$\Delta H_f^\circ = -308.2$ kcal $\Delta H_f^\circ = -311.3$ kcal

Table 4-3

Enthalpies and Entropies of Weak Acids and Their Anions

	ΔH_f°, kcal/mole	S°, eu
H_2SO_3	-145.5	56
HSO_3^-	-151.9	26
SO_3^{2-}	-151.9	-7
HSO_4^-	-211.7	30.3
SO_4^{2-}	-216.9	4.1
H_3PO_4	-308.2	42.1
$H_2PO_4^-$	-311.3	21.3
HPO_4^{2-}	-310.4	-8.6
PO_4^{3-}	-306.9	-52
HCO_3^-	-165.2	22.7
CO_3^{2-}	-161.6	-12.7
$HCOOH$	-98.0	39.1
$HCOO^-$	-98.0	21.9
$HClO_2$	-13.68	42
ClO_2^-	-17.18	24.1
$HClO$	-27.8	31
ClO^-	-23.9	10

The entropy and energy of the aqueous proton are zero, by definition. These data give us $\Delta H^\circ = -3.1$ kcal and $\Delta S^\circ = -20.8$ eu for the dissociation. The reaction is exothermic, and the entropy decreases! While these results are just the opposite of our naïve expectation, a little reflection provides a satisfying explanation of the observation. In the ionization reaction, a proton is not really set free from a confined position on H_3PO_4, but is only *transferred* from H_3PO_4 to a solvent molecule to form a hydrated H_3O^+ ion. That is, although the reaction looks like a dissociation, it is actually only a relocation of the proton, and this does not increase the entropy of the system.

But why does the entropy decrease? The ionization does "create" charge, and water molecules will tend to be strongly bound to each of the charge centers, owing to the action of the ionic charge on the dipole moment of the water molecule, as shown in Fig. 4–9. This charge-induced hydration binds or localizes water molecules which would otherwise be more free to assume a variety of random positions in the solvent. In effect, by binding the water molecules, the ions tend to create order in the solvent; and this effect is responsible for the decrease in entropy which accompanies ionization.

Naturally, it is also this hydration of ions which supplies the energy necessary to remove the proton from the acid. In the case of the first ionization of H_3PO_4, this hydration energy is sufficient to overcome the energy required to separate a proton from the parent acid, and the reaction is exothermic. To see the relative importance of the entropy and enthalpy effects, we evaluate the ionization constant K_1. At 298°K we find

$$K_1 = 10^{\Delta S°/2.3R} \times 10^{-\Delta H°/2.3RT} = 10^{-4.5} \times 10^{2.3} = 6 \times 10^{-3}$$

The entropy effect is larger than the enthalpy factor, and the first dissociation constant is less than unity.

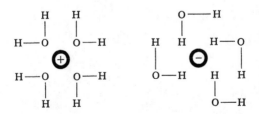

Figure 4–9 The presence of ions localizes and orients water molecules, reducing their freedom and entropy.

Let us apply a similar analysis to the second ionization constant of phosphoric acid.

$$H_2PO_4^- = H^+ + HPO_4^-$$
$$\Delta H = 0.9 \text{ kcal} \qquad \Delta S° = -29.9 \text{ eu}$$
$$K_2 = 10^{-6.5} \times 10^{-0.66} = 7 \times 10^{-7}$$

This ionization reaction is slightly endothermic, and again the entropy of ionization is negative.

For the third ionization

$$HPO_4^- = H^+ + PO_4^{3-}$$

Table 4–3 yields $\Delta H° = 4.5$ kcal, $\Delta S° = -43.3$ eu, and

$$K_3 = 10^{-9.5} \times 10^{-2.5} = 10^{-12}$$

Comparison of these three ionizations shows, as we might have expected, that more energy is required to remove a proton as the negative charge on the parent acid increases. However, the data also show that entropy is just as important a factor as enthalpy in causing the second and third ionization constants to be smaller than the first.

Another particularly good example of the importance of entropy in the determination of acid strength involves acetic acid and trichloroacetic acid, which ionize according to the equations

acetic acid $K_a = 1.8 \times 10^{-5}$

trichloroacetic
acid $K_a = 3 \times 10^{-1}$

Despite the fact that the ionization constant for trichloroacetic acid is nearly 20,000 times that of acetic acid, the enthalpy change for both these ionizations is approximately zero. The difference in acid strength must arise entirely from entropy effects. Experimental measurements show that $\Delta S° = -22$ eu for the ionization of acetic acid and $\Delta S° = -2$ eu for the ionization of trichloroacetic acid. Clearly, the greater entropy loss upon ionization makes acetic acid weaker than trichloroacetic acid.

Electrode Reactions

Thermodynamic analysis is also helpful in understanding oxidation-reduction reactions. For example, the reaction

$$Zn(s) + Cu^{2+}(aq) = Zn^{2+}(aq) + Cu(s)$$

has a standard potential of 1.10 volts, or a standard free energy change of -50.7 kcal, associated with it. While these numbers make it clear that the equilibrium state favors the products, we might ask what factor really provides the driving force of the reaction. Is it entropy or enthalpy? Tables 2–2 and 3–1 give us $\Delta H° = -51.8$ kcal, $\Delta S° = -3.18$ eu, and

$$K = 10^{-0.86} \times 10^{38.1} = 2 \times 10^{37}$$

at 298°K. Thus the entropy change is very small and tends to decrease the equilibrium constant, while the enthalpy change is entirely responsible for the large equilibrium constant.

Now that we are assured that it is the enthalpy change which causes zinc metal to reduce cupric ion so completely, we can seek the features of the reactants and products which cause the enthalpy to change as it does. One way to find the answer is to split the reaction into two half-reactions:

$$Zn(s) = Zn^{2+}(aq) + 2e$$
$$2e + Cu^{2+}(aq) = Cu(s)$$

Now let us give our attention to the first half-reaction and further decompose it into a series of steps:

$$
\begin{array}{ll}
\mathrm{Zn}(s) = \mathrm{Zn}(g) & \Delta H = \Delta H_{\mathrm{vap}} = 31.2 \text{ kcal} \\
\mathrm{Zn}(g) = \mathrm{Zn^{2+}}(g) + 2e & \Delta H = I \quad\quad = 630 \text{ kcal} \\
\underline{\mathrm{Zn^{2+}}(g) = \mathrm{Zn^{2+}}(aq)} & \underline{\Delta H = \Delta H_h \quad = -490} \\
\mathrm{Zn}(s) = \mathrm{Zn^{2+}}(aq) + 2e & \Delta H \quad\quad\quad = 171 \text{ kcal}
\end{array}
$$

The first step is the vaporization of zinc metal to gaseous zinc atoms, and the second step is the ionization of the gaseous zinc atoms to form gaseous zinc ions and two electrons. The enthalpy changes for these two processes are known very accurately. The third step is the hydration of the gaseous ion to form the aqueous ion, and the enthalpy change, called the enthalpy of hydration, can only be estimated.

The copper half-reaction can also be similarly written as the sum of three steps.

$$
\begin{array}{ll}
\mathrm{Cu}(g) = \mathrm{Cu}(g) & \Delta H = 81.1 \text{ kcal} \\
\mathrm{Cu}(g) = \mathrm{Cu^{2+}}(g) + 2e & \Delta H = 645 \\
\underline{\mathrm{Cu^{2+}}(g) = \mathrm{Cu^{2+}}(aq)} & \underline{\Delta H = -510} \\
\mathrm{Cu}(s) = \mathrm{Cu^{2+}}(aq) + 2e & \Delta H = 220
\end{array}
$$

This trick of separating the reaction into several steps is rather artificial, since at no time during the actual oxidation-reduction process do gaseous ions or free electrons actually exist. However, we can find the principal reason for the magnitude of the enthalpy change of the reaction if we recombine the steps pairwise in the following manner:

$$
\begin{array}{lll}
\mathrm{Zn}(s) = \mathrm{Zn}(g) & \Delta H = \left.\begin{array}{r} 31.2 \\ \end{array}\right\} & \\
\mathrm{Cu}(g) = \mathrm{Cu}(s) & \Delta H = \left.\begin{array}{r} -81.1 \\ \end{array}\right\} & \Delta H = -49.9 \\
\mathrm{Zn}(g) = \mathrm{Zn^{2+}}(g) + 2e & \Delta H = \left.\begin{array}{r} 630 \\ \end{array}\right\} & \\
2e + \mathrm{Cu^{2+}}(g) = \mathrm{Cu}(g) & \Delta H = \left.\begin{array}{r} -645 \\ \end{array}\right\} & \Delta H = -15 \\
\mathrm{Zn^{2+}}(g) = \mathrm{Zn^{2+}}(aq) & \Delta H = \left.\begin{array}{r} -490 \\ \end{array}\right\} & \\
\underline{\mathrm{Cu^{2+}}(aq) = \mathrm{Cu^{2+}}(g)} & \underline{\Delta H = \left.\begin{array}{r} 510 \\ \end{array}\right\}} & \underline{\Delta H = 20} \\
\mathrm{Zn}(s) + \mathrm{Cu^{2+}}(aq) = \mathrm{Zn^{2+}}(aq) + \mathrm{Cu}(s) & \Delta H = -45 \text{ kcal}
\end{array}
$$

Since the hydration energies are only estimated, the ΔH calculated by this method does not agree exactly with the measured value of -50.1 kcal. However, by inspecting the values of ΔH for each of the pairs of steps, it becomes clear that the principal reason the reaction is exothermic is that there is a 49.9-kcal difference between the vaporization enthalpies of zinc and copper. The enthalpy of vaporization measures the energy required to separate the atoms in a crystal. Therefore, we can say that the most important cause of the exothermicity is that in copper metal the atoms are more tightly bound to each other than are zinc atoms in zinc metal. The other factor which tends to make the reaction exothermic is that the gaseous copper atom binds its two electrons with more energy than does a gaseous zinc atom. These effects are only slightly offset by the fact that the hydration enthalpy of cupric ion is more negative than that of zinc ion.

This analysis of the reaction between zinc and cupric ion shows how thermodynamic reasoning can be used to expose the molecular factors which control a complex chemical phenomenon. Of course, questions such as why copper has a larger enthalpy of vaporization than zinc are left unanswered, since these are problems in electronic structure. Thermodynamics answers no questions by itself, but it does show how one experimental property is related to another.

4-4 HEAT ENGINES

The early workers in thermodynamics were very much concerned with the operation and efficiency of devices for converting heat into useful work. In fact, there are two common formulations of the second law of thermodynamics which are stated in terms of the possibility of converting heat into work. In this section we shall show how the principle of entropy increase limits the efficiency of the continuous conversion of heat to work. To do so, we shall analyze the behavior of an idealized device called the Carnot heat engine.

The Carnot engine consists of a source of heat that is maintained at a constant high temperature T_h. In an actual engine this source of heat might be a furnace or nuclear reactor. In addition, there is a heat sink, or cold reservoir, maintained at a low temperature T_c. This heat sink receives any heat which might be discarded by the engine as it operates. In a working engine this cold reservoir might be the atmosphere or a cooling bath. The engine operates by carrying a working substance through a cycle, or a sequence of changes which converts heat into work. At the end of each cycle, the working substance has returned to its original state and a new cycle begins. The working material is most often a gas, but in principle it may be any substance. For convenience in the following discussion, we shall refer to the working substance as a gas.

The cycle used in the Carnot engine is shown in Fig. 4–10, and the accompanying changes of state of the working gas are shown in Fig. 4–11. There are four steps, *each of them carried out reversibly:*

1. In the first step the working gas absorbs q_1 units of heat from the hot reservoir; while doing so, the gas expands isothermally and does work w_1 on the outside world. Notice that since the gas absorbs heat, q_1 is a positive number.

2. By means of heat insulation, the working substance is thermally isolated from its surroundings, but the mechanical link with the outside world is maintained. The gas expands, doing work w_2. Since it is thermally isolated, it absorbs no heat during this step, and $q_2 = 0$. Such a process is called an *adiabatic expansion.* Since the gas does work without receiving heat, its internal energy and temperature decrease. The adiabatic expansion continues until the temperature of the gas has dropped from T_h to T_c, the temperature of the cold reservoir.

3. The gas is brought into contact with the cold reservoir and then compressed isothermally. This process deposits heat into the cold reservoir, so q_3, *the heat absorbed by the gas,* is a negative number. Work is done on the gas as it is compressed, so w_3 is also a negative number.

4. In the final step of the cycle, the gas is again thermally insulated and compressed adiabatically. Therefore, $q_4 = 0$ and w_4 is a negative number. During the compression, the gas temperature rises to T_h.

At the end of the cycle the gas has returned to its initial state. Therefore, $\Delta S_{gas} = 0$, $\Delta E_{gas} = 0$, and

$$q_1 + q_3 = w_1 + w_2 + w_3 + w_4 = w$$

where w is the net work done by the gas during the cycle.

Now we must consider the entropy changes that occur during one cycle. According to the second law of thermodynamics, the

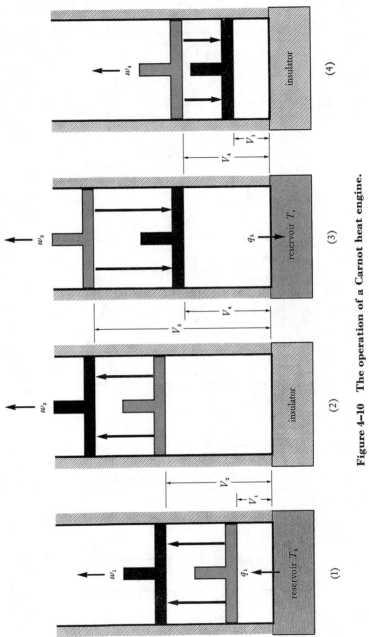

Figure 4–10 The operation of a Carnot heat engine.

(a)

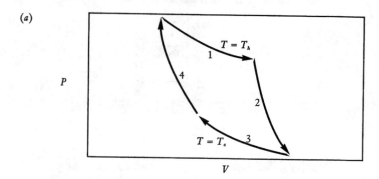

(b)

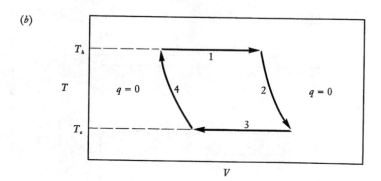

(c)

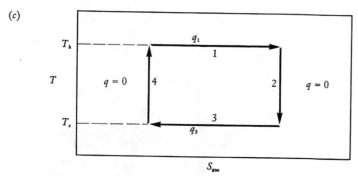

Figure 4–11 Alternative representations of the steps involved in the Carnot cycle.

total entropy change of the gas and the two reservoirs must be zero if the whole cycle is carried out reversibly. That is,

$$\Delta S_{gas} + \Delta S_{res} = 0 \qquad \text{(reversible cycle)}$$

But $\Delta S_{gas} = 0$; therefore, $\Delta S_{res} = 0$, since we have said the cycle is carried out reversibly. In step 1, the hot reservoir loses q_1 units of heat, so its entropy change is $-q_1/T_h$. In step 3, the cold reservoir gains $|q_3|$ units of heat. The entropy change of a reservoir which gains $|q_3|$ units of heat is $|q_3|/T$ or $-q_3/T_c$, since q_3 is a negative number. Therefore, we can write

$$\Delta S_{res} = 0 \qquad \text{(reversible cycle)}$$

$$-\frac{q_1}{T_h} - \frac{q_3}{T_c} = 0$$

But

$$q_1 + q_3 = w$$

On combining these equations, we find

$$-\frac{q_1}{T_h} + \frac{q_1}{T_c} - \frac{w}{T_c} = 0 \qquad \text{(reversible cycle)}$$

$$\frac{w}{q_1} = 1 - \frac{T_c}{T_h} \equiv \eta \qquad (4\text{-}21)$$

The quantity w/q_1 is just the net work done on the outside world divided by the heat extracted from the high-temperature source. Therefore, it is the fraction of the heat q_1 which is converted to work, and it is called the efficiency η of the reversible Carnot engine.

Equation (4-21) has two important aspects. First, it shows that in order for the efficiency of the engine to approach unity, T_h must approach infinity, or T_c must approach zero. The efficiency of even a perfectly frictionless engine is always less than unity, as long as the engine operates between any two finite temperatures. This fact is a direct consequence of the second law of thermodynamics, which requires that as the engine operates reversibly, the entropy of the universe must stay constant. Since the working substance of the engine returns to its original state at the end of each cycle, its entropy stays constant. Therefore, the entropy of the reservoirs must remain constant. In the first step of the cycle, the hot reservoir loses heat q_1, and its entropy decreases by

$-q_1/T_h$. If the engine had unit efficiency, q_1 units of heat would be completely converted to work, the entropy of the universe would decrease, and the second law of thermodynamics would be violated. If, however, some of the heat q_1 is discarded to the cold reservoir in step 3, the entropy increase, q_3/T_c, of this cold reservoir can completely compensate for the entropy decrease of the hot reservoir. Therefore, since some of the heat q_1 must be discarded, it cannot all be converted to work, and the engine must have less than unit efficiency.

The second important feature of Eq. (4–21) is that it shows us how to define a temperature scale which is completely independent of the particular properties of any substance. In order to measure the ratio of any two temperatures T_c and T_h, we need only start a reversible Carnot engine operating between these temperatures and measure its efficiency η. The size of the degree can be fixed by arbitrarily taking the temperature of melting ice to be 273.15°. This procedure defines the absolute, or thermodynamic, temperature scale. Since we made no mention of the properties of the working substance in the derivation of Eq. (4–21), the thermodynamic temperature scale is completely independent of the properties of any substance.

Measurement of temperature using a Carnot engine could be extremely inconvenient and is never done in practice. However, by imagining that we use an ideal gas as the working substance of a Carnot engine, it is not difficult to prove that the thermodynamic temperature scale and the ideal-gas temperature scale are identical. Therefore, the ideal-gas thermometer is merely a convenient way to measure the thermodynamic temperature.

4–5 CONCLUSION

The laws of thermodynamics, when stated briefly, are highly abstract. They are so because they are designed to state what is universally common to all our varied experience concerning the behavior of macroscopic systems at equilibrium. Owing to this broad experimental basis, they stand as principles of unsurpassed generality and as guideposts of unmatched reliability for the investigation and understanding of equilibrium phenomena.

Despite the abstract nature of the laws, their application can be of a highly specific and practical nature. To the chemical engineer they give a scheme by which he can calculate the optimum conditions for, and maximum yield of, a chemical conversion. From the thermodynamic laws the chemist forms a quantitative language for specifying the "stability" and correlating the properties of chemical substances. The problems which face the theoretical chemist are stated in thermodynamic language. A metal shows an extraordinarily large entropy at low temperature; what facet of atomic or electronic behavior is responsible? Of the halogen molecules, fluorine requires the least energy to dissociate it into atoms, while its neighbor chlorine requires the most. Can our theories of electronic structure account for this? We have tried to illustrate how the thermodynamic laws lead to these applications. If the preceding pages have shown that the thermodynamic laws are not only an elegant statement of what has been learned but also an eminently practical guide to the understanding of new chemical phenomena, our purpose has been accomplished.

SUPPLEMENTARY READING

Slightly more advanced treatments of chemical thermodynamics are to be found in textbooks of physical chemistry. The following are three of the most readable.

Daniels, F., and R. A. Alberty: "Physical Chemistry," Wiley, New York, 1961.

Hamill, W. H., and R. R. Williams: "Principles of Physical Chemistry," Prentice-Hall, Englewood Cliffs, N.J., 1959.

Moore, W. J.: "Physical Chemistry," Prentice-Hall, Englewood Cliffs, N.J., 1962.

These textbooks of chemical thermodynamics are notable for their clarity.

Caldin, E. F.: "Introduction to Chemical Thermodynamics," Oxford, London, 1958.

Fermi, E.: "Thermodynamics," Dover, New York, 1956.

Wall, F. T.: "Chemical Thermodynamics," Freeman, San Francisco, 1958.

PROBLEMS

1. The standard free energy change for the vaporization of molten aluminum at 1860°K is 12.15 kcal/mole. What is the vapor pressure of aluminum at this temperature?

2. The vapor pressure of liquid sulfur dioxide at four temperatures is given below. From these data determine ΔH_{vap} by a graphical method. Is ΔH_{vap} constant over the whole temperature range?

P, mm	760	400	100	40
T, °K	263	251	226	213

3. The molar volume of liquid mercury is 14.653 ml, and that of solid mercury is 14.133 ml at the melting point, 234°K. If the enthalpy of fusion is 565 cal/mole, what is the change in the melting point produced by the application of 100 atm pressure?

4. Five grams of a polymer having a molecular weight of 200,000 is dissolved in 1000 grams (1140 ml) of benzene at 300°K. Calculate the boiling point elevation, the freezing point depression, the change in the vapor pressure of the benzene, and the osmotic pressure. The vapor pressure of benzene is 100 mm at 300°K.

5. Equation (4–18) is the expression for the solubility of a solute in an ideal solution. Evaluate the constant of integration for a solution of naphthalene in benzene from the facts that for pure naphthalene dissolving in itself (that is, melting), $x_2 = 1$ at $T = 80°C$ and $\Delta H^\circ_{fus} = 4600$ cal/mole. Use this constant to calculate the solubility of naphthalene in benzene at 25°C.

6. The reactions listed below can be combined to form an alternate path for the reaction $HX(aq) = H^+(aq) + X^-(aq)$, where X stands for F or I. From the data given calculate ΔH° for the ionization of aqueous HF and HI. By comparing the data for each step of the ionization process, decide why the ionization of HI is more exothermic than that of HF.

	ΔH(HF), kcal	ΔH(HI), kcal
$HX(aq) = HX(g)$	−11.5	−5.5
$HX(g) = H(g) + X(g)$	134.6	71.4
$H(g) + X(g) = H^+(g) + X^-(g)$	232.8	239.3
$H^+(g) + X^-(g) = H^+(aq) + X^-(aq)$	−381.9	−330.3

Index

A CATALOG OF SELECTED
DOVER BOOKS
IN SCIENCE AND MATHEMATICS

Astronomy

BURNHAM'S CELESTIAL HANDBOOK, Robert Burnham, Jr. Thorough guide to the stars beyond our solar system. Exhaustive treatment. Alphabetical by constellation: Andromeda to Cetus in Vol. 1; Chamaeleon to Orion in Vol. 2; and Pavo to Vulpecula in Vol. 3. Hundreds of illustrations. Index in Vol. 3. 2,000pp. 6⅛ x 9¼.
Vol. I: 0-486-23567-X
Vol. II: 0-486-23568-8
Vol. III: 0-486-23673-0

EXPLORING THE MOON THROUGH BINOCULARS AND SMALL TELESCOPES, Ernest H. Cherrington, Jr. Informative, profusely illustrated guide to locating and identifying craters, rills, seas, mountains, other lunar features. Newly revised and updated with special section of new photos. Over 100 photos and diagrams. 240pp. 8¼ x 11. 0-486-24491-1

THE EXTRATERRESTRIAL LIFE DEBATE, 1750–1900, Michael J. Crowe. First detailed, scholarly study in English of the many ideas that developed from 1750 to 1900 regarding the existence of intelligent extraterrestrial life. Examines ideas of Kant, Herschel, Voltaire, Percival Lowell, many other scientists and thinkers. 16 illustrations. 704pp. 5⅜ x 8½. 0-486-40675-X

THEORIES OF THE WORLD FROM ANTIQUITY TO THE COPERNICAN REVOLUTION, Michael J. Crowe. Newly revised edition of an accessible, enlightening book recreates the change from an earth-centered to a sun-centered conception of the solar system. 242pp. 5⅜ x 8½. 0-486-41444-2

A HISTORY OF ASTRONOMY, A. Pannekoek. Well-balanced, carefully reasoned study covers such topics as Ptolemaic theory, work of Copernicus, Kepler, Newton, Eddington's work on stars, much more. Illustrated. References. 521pp. 5⅜ x 8½.
0-486-65994-1

A COMPLETE MANUAL OF AMATEUR ASTRONOMY: TOOLS AND TECHNIQUES FOR ASTRONOMICAL OBSERVATIONS, P. Clay Sherrod with Thomas L. Koed. Concise, highly readable book discusses: selecting, setting up and maintaining a telescope; amateur studies of the sun; lunar topography and occultations; observations of Mars, Jupiter, Saturn, the minor planets and the stars; an introduction to photoelectric photometry; more. 1981 ed. 124 figures. 25 halftones. 37 tables. 335pp. 6½ x 9¼. 0-486-40675-X

AMATEUR ASTRONOMER'S HANDBOOK, J. B. Sidgwick. Timeless, comprehensive coverage of telescopes, mirrors, lenses, mountings, telescope drives, micrometers, spectroscopes, more. 189 illustrations. 576pp. 5⅝ x 8¼. (Available in U.S. only.)
0-486-24034-7

STARS AND RELATIVITY, Ya. B. Zel'dovich and I. D. Novikov. Vol. 1 of *Relativistic Astrophysics* by famed Russian scientists. General relativity, properties of matter under astrophysical conditions, stars, and stellar systems. Deep physical insights, clear presentation. 1971 edition. References. 544pp. 5⅜ x 8¼. 0-486-69424-0

Chemistry

THE SCEPTICAL CHYMIST: THE CLASSIC 1661 TEXT, Robert Boyle. Boyle defines the term "element," asserting that all natural phenomena can be explained by the motion and organization of primary particles. 1911 ed. viii+232pp. 5⅜ x 8½.
0-486-42825-7

RADIOACTIVE SUBSTANCES, Marie Curie. Here is the celebrated scientist's doctoral thesis, the prelude to her receipt of the 1903 Nobel Prize. Curie discusses establishing atomic character of radioactivity found in compounds of uranium and thorium; extraction from pitchblende of polonium and radium; isolation of pure radium chloride; determination of atomic weight of radium; plus electric, photographic, luminous, heat, color effects of radioactivity. ii+94pp. 5⅜ x 8½.   0-486-42550-9

CHEMICAL MAGIC, Leonard A. Ford. Second Edition, Revised by E. Winston Grundmeier. Over 100 unusual stunts demonstrating cold fire, dust explosions, much more. Text explains scientific principles and stresses safety precautions. 128pp. 5⅜ x 8½.   0-486-67628-5

THE DEVELOPMENT OF MODERN CHEMISTRY, Aaron J. Ihde. Authoritative history of chemistry from ancient Greek theory to 20th-century innovation. Covers major chemists and their discoveries. 209 illustrations. 14 tables. Bibliographies. Indices. Appendices. 851pp. 5⅜ x 8½.   0-486-64235-6

CATALYSIS IN CHEMISTRY AND ENZYMOLOGY, William P. Jencks. Exceptionally clear coverage of mechanisms for catalysis, forces in aqueous solution, carbonyl- and acyl-group reactions, practical kinetics, more. 864pp. 5⅜ x 8½.
0-486-65460-5

ELEMENTS OF CHEMISTRY, Antoine Lavoisier. Monumental classic by founder of modern chemistry in remarkable reprint of rare 1790 Kerr translation. A must for every student of chemistry or the history of science. 539pp. 5⅜ x 8½. 0-486-64624-6

THE HISTORICAL BACKGROUND OF CHEMISTRY, Henry M. Leicester. Evolution of ideas, not individual biography. Concentrates on formulation of a coherent set of chemical laws. 260pp. 5⅜ x 8½.   0-486-61053-5

A SHORT HISTORY OF CHEMISTRY, J. R. Partington. Classic exposition explores origins of chemistry, alchemy, early medical chemistry, nature of atmosphere, theory of valency, laws and structure of atomic theory, much more. 428pp. 5⅜ x 8½. (Available in U.S. only.)   0-486-65977-1

GENERAL CHEMISTRY, Linus Pauling. Revised 3rd edition of classic first-year text by Nobel laureate. Atomic and molecular structure, quantum mechanics, statistical mechanics, thermodynamics correlated with descriptive chemistry. Problems. 992pp. 5⅜ x 8½.   0-486-65622-5

FROM ALCHEMY TO CHEMISTRY, John Read. Broad, humanistic treatment focuses on great figures of chemistry and ideas that revolutionized the science. 50 illustrations. 240pp. 5⅜ x 8½.   0-486-28690-8

Engineering

DE RE METALLICA, Georgius Agricola. The famous Hoover translation of greatest treatise on technological chemistry, engineering, geology, mining of early modern times (1556). All 289 original woodcuts. 638pp. 6¾ x 11. 0-486-60006-8

FUNDAMENTALS OF ASTRODYNAMICS, Roger Bate et al. Modern approach developed by U.S. Air Force Academy. Designed as a first course. Problems, exercises. Numerous illustrations. 455pp. 5⅜ x 8½. 0-486-60061-0

DYNAMICS OF FLUIDS IN POROUS MEDIA, Jacob Bear. For advanced students of ground water hydrology, soil mechanics and physics, drainage and irrigation engineering and more. 335 illustrations. Exercises, with answers. 784pp. 6⅛ x 9¼. 0-486-65675-6

THEORY OF VISCOELASTICITY (Second Edition), Richard M. Christensen. Complete consistent description of the linear theory of the viscoelastic behavior of materials. Problem-solving techniques discussed. 1982 edition. 29 figures. xiv+364pp. 6⅛ x 9¼. 0-486-42880-X

MECHANICS, J. P. Den Hartog. A classic introductory text or refresher. Hundreds of applications and design problems illuminate fundamentals of trusses, loaded beams and cables, etc. 334 answered problems. 462pp. 5⅜ x 8½. 0-486-60754-2

MECHANICAL VIBRATIONS, J. P. Den Hartog. Classic textbook offers lucid explanations and illustrative models, applying theories of vibrations to a variety of practical industrial engineering problems. Numerous figures. 233 problems, solutions. Appendix. Index. Preface. 436pp. 5⅜ x 8½. 0-486-64785-4

STRENGTH OF MATERIALS, J. P. Den Hartog. Full, clear treatment of basic material (tension, torsion, bending, etc.) plus advanced material on engineering methods, applications. 350 answered problems. 323pp. 5⅜ x 8½. 0-486-60755-0

A HISTORY OF MECHANICS, René Dugas. Monumental study of mechanical principles from antiquity to quantum mechanics. Contributions of ancient Greeks, Galileo, Leonardo, Kepler, Lagrange, many others. 671pp. 5⅜ x 8½. 0-486-65632-2

STABILITY THEORY AND ITS APPLICATIONS TO STRUCTURAL MECHANICS, Clive L. Dym. Self-contained text focuses on Koiter postbuckling analyses, with mathematical notions of stability of motion. Basing minimum energy principles for static stability upon dynamic concepts of stability of motion, it develops asymptotic buckling and postbuckling analyses from potential energy considerations, with applications to columns, plates, and arches. 1974 ed. 208pp. 5⅜ x 8½.
0-486-42541-X

METAL FATIGUE, N. E. Frost, K. J. Marsh, and L. P. Pook. Definitive, clearly written, and well-illustrated volume addresses all aspects of the subject, from the historical development of understanding metal fatigue to vital concepts of the cyclic stress that causes a crack to grow. Includes 7 appendixes. 544pp. 5⅜ x 8½. 0-486-40927-9

ROCKETS, Robert Goddard. Two of the most significant publications in the history of rocketry and jet propulsion: "A Method of Reaching Extreme Altitudes" (1919) and "Liquid Propellant Rocket Development" (1936). 128pp. 5⅜ x 8½.   0-486-42537-1

STATISTICAL MECHANICS: PRINCIPLES AND APPLICATIONS, Terrell L. Hill. Standard text covers fundamentals of statistical mechanics, applications to fluctuation theory, imperfect gases, distribution functions, more. 448pp. 5⅜ x 8½.

0-486-65390-0

ENGINEERING AND TECHNOLOGY 1650–1750: ILLUSTRATIONS AND TEXTS FROM ORIGINAL SOURCES, Martin Jensen. Highly readable text with more than 200 contemporary drawings and detailed engravings of engineering projects dealing with surveying, leveling, materials, hand tools, lifting equipment, transport and erection, piling, bailing, water supply, hydraulic engineering, and more. Among the specific projects outlined-transporting a 50-ton stone to the Louvre, erecting an obelisk, building timber locks, and dredging canals. 207pp. 8⅜ x 11¼.

0-486-42232-1

THE VARIATIONAL PRINCIPLES OF MECHANICS, Cornelius Lanczos. Graduate level coverage of calculus of variations, equations of motion, relativistic mechanics, more. First inexpensive paperbound edition of classic treatise. Index. Bibliography. 418pp. 5⅜ x 8½.   0-486-65067-7

PROTECTION OF ELECTRONIC CIRCUITS FROM OVERVOLTAGES, Ronald B. Standler. Five-part treatment presents practical rules and strategies for circuits designed to protect electronic systems from damage by transient overvoltages. 1989 ed. xxiv+434pp. 6⅛ x 9¼.   0-486-42552-5

ROTARY WING AERODYNAMICS, W. Z. Stepniewski. Clear, concise text covers aerodynamic phenomena of the rotor and offers guidelines for helicopter performance evaluation. Originally prepared for NASA. 537 figures. 640pp. 6⅛ x 9¼.

0-486-64647-5

INTRODUCTION TO SPACE DYNAMICS, William Tyrrell Thomson. Comprehensive, classic introduction to space-flight engineering for advanced undergraduate and graduate students. Includes vector algebra, kinematics, transformation of coordinates. Bibliography. Index. 352pp. 5⅜ x 8½.   0-486-65113-4

HISTORY OF STRENGTH OF MATERIALS, Stephen P. Timoshenko. Excellent historical survey of the strength of materials with many references to the theories of elasticity and structure. 245 figures. 452pp. 5⅜ x 8½.   0-486-61187-6

ANALYTICAL FRACTURE MECHANICS, David J. Unger. Self-contained text supplements standard fracture mechanics texts by focusing on analytical methods for determining crack-tip stress and strain fields. 336pp. 6⅛ x 9¼.   0-486-41737-9

STATISTICAL MECHANICS OF ELASTICITY, J. H. Weiner. Advanced, self-contained treatment illustrates general principles and elastic behavior of solids. Part 1, based on classical mechanics, studies thermoelastic behavior of crystalline and polymeric solids. Part 2, based on quantum mechanics, focuses on interatomic force laws, behavior of solids, and thermally activated processes. For students of physics and chemistry and for polymer physicists. 1983 ed. 96 figures. 496pp. 5⅜ x 8½.

0-486-42260-7

Physics

OPTICAL RESONANCE AND TWO-LEVEL ATOMS, L. Allen and J. H. Eberly. Clear, comprehensive introduction to basic principles behind all quantum optical resonance phenomena. 53 illustrations. Preface. Index. 256pp. 5⅜ x 8½. 0-486-65533-4

QUANTUM THEORY, David Bohm. This advanced undergraduate-level text presents the quantum theory in terms of qualitative and imaginative concepts, followed by specific applications worked out in mathematical detail. Preface. Index. 655pp. 5⅜ x 8½. 0-486-65969-0

ATOMIC PHYSICS (8th EDITION), Max Born. Nobel laureate's lucid treatment of kinetic theory of gases, elementary particles, nuclear atom, wave-corpuscles, atomic structure and spectral lines, much more. Over 40 appendices, bibliography. 495pp. 5⅜ x 8½. 0-486-65984-4

A SOPHISTICATE'S PRIMER OF RELATIVITY, P. W. Bridgman. Geared toward readers already acquainted with special relativity, this book transcends the view of theory as a working tool to answer natural questions: What is a frame of reference? What is a "law of nature"? What is the role of the "observer"? Extensive treatment, written in terms accessible to those without a scientific background. 1983 ed. xlviii+172pp. 5⅜ x 8½. 0-486-42549-5

AN INTRODUCTION TO HAMILTONIAN OPTICS, H. A. Buchdahl. Detailed account of the Hamiltonian treatment of aberration theory in geometrical optics. Many classes of optical systems defined in terms of the symmetries they possess. Problems with detailed solutions. 1970 edition. xv + 360pp. 5⅜ x 8½. 0-486-67597-1

PRIMER OF QUANTUM MECHANICS, Marvin Chester. Introductory text examines the classical quantum bead on a track: its state and representations; operator eigenvalues; harmonic oscillator and bound bead in a symmetric force field; and bead in a spherical shell. Other topics include spin, matrices, and the structure of quantum mechanics; the simplest atom; indistinguishable particles; and stationary-state perturbation theory. 1992 ed. xiv+314pp. 6⅛ x 9¼. 0-486-42878-8

LECTURES ON QUANTUM MECHANICS, Paul A. M. Dirac. Four concise, brilliant lectures on mathematical methods in quantum mechanics from Nobel Prize-winning quantum pioneer build on idea of visualizing quantum theory through the use of classical mechanics. 96pp. 5⅜ x 8½. 0-486-41713-1

THIRTY YEARS THAT SHOOK PHYSICS: THE STORY OF QUANTUM THEORY, George Gamow. Lucid, accessible introduction to influential theory of energy and matter. Careful explanations of Dirac's anti-particles, Bohr's model of the atom, much more. 12 plates. Numerous drawings. 240pp. 5⅜ x 8½. 0-486-24895-X

ELECTRONIC STRUCTURE AND THE PROPERTIES OF SOLIDS: THE PHYSICS OF THE CHEMICAL BOND, Walter A. Harrison. Innovative text offers basic understanding of the electronic structure of covalent and ionic solids, simple metals, transition metals and their compounds. Problems. 1980 edition. 582pp. 6⅛ x 9¼. 0-486-66021-4

HYDRODYNAMIC AND HYDROMAGNETIC STABILITY, S. Chandrasekhar. Lucid examination of the Rayleigh-Benard problem; clear coverage of the theory of instabilities causing convection. 704pp. 5⅝ x 8¼. 0-486-64071-X

INVESTIGATIONS ON THE THEORY OF THE BROWNIAN MOVEMENT, Albert Einstein. Five papers (1905–8) investigating dynamics of Brownian motion and evolving elementary theory. Notes by R. Fürth. 122pp. 5⅜ x 8½. 0-486-60304-0

THE PHYSICS OF WAVES, William C. Elmore and Mark A. Heald. Unique overview of classical wave theory. Acoustics, optics, electromagnetic radiation, more. Ideal as classroom text or for self-study. Problems. 477pp. 5⅜ x 8½. 0-486-64926-1

GRAVITY, George Gamow. Distinguished physicist and teacher takes reader-friendly look at three scientists whose work unlocked many of the mysteries behind the laws of physics: Galileo, Newton, and Einstein. Most of the book focuses on Newton's ideas, with a concluding chapter on post-Einsteinian speculations concerning the relationship between gravity and other physical phenomena. 160pp. 5⅜ x 8½. 0-486-42563-0

PHYSICAL PRINCIPLES OF THE QUANTUM THEORY, Werner Heisenberg. Nobel Laureate discusses quantum theory, uncertainty, wave mechanics, work of Dirac, Schroedinger, Compton, Wilson, Einstein, etc. 184pp. 5⅜ x 8½. 0-486-60113-7

ATOMIC SPECTRA AND ATOMIC STRUCTURE, Gerhard Herzberg. One of best introductions; especially for specialist in other fields. Treatment is physical rather than mathematical. 80 illustrations. 257pp. 5⅜ x 8½. 0-486-60115-3

AN INTRODUCTION TO STATISTICAL THERMODYNAMICS, Terrell L. Hill. Excellent basic text offers wide-ranging coverage of quantum statistical mechanics, systems of interacting molecules, quantum statistics, more. 523pp. 5⅜ x 8½. 0-486-65242-4

THEORETICAL PHYSICS, Georg Joos, with Ira M. Freeman. Classic overview covers essential math, mechanics, electromagnetic theory, thermodynamics, quantum mechanics, nuclear physics, other topics. First paperback edition. xxiii + 885pp. 5⅜ x 8½. 0-486-65227-0

PROBLEMS AND SOLUTIONS IN QUANTUM CHEMISTRY AND PHYSICS, Charles S. Johnson, Jr. and Lee G. Pedersen. Unusually varied problems, detailed solutions in coverage of quantum mechanics, wave mechanics, angular momentum, molecular spectroscopy, more. 280 problems plus 139 supplementary exercises. 430pp. 6½ x 9¼. 0-486-65236-X

THEORETICAL SOLID STATE PHYSICS, Vol. 1: Perfect Lattices in Equilibrium; Vol. II: Non-Equilibrium and Disorder, William Jones and Norman H. March. Monumental reference work covers fundamental theory of equilibrium properties of perfect crystalline solids, non-equilibrium properties, defects and disordered systems. Appendices. Problems. Preface. Diagrams. Index. Bibliography. Total of 1,301pp. 5⅜ x 8½. Two volumes. Vol. I: 0-486-65015-4 Vol. II: 0-486-65016-2

WHAT IS RELATIVITY? L. D. Landau and G. B. Rumer. Written by a Nobel Prize physicist and his distinguished colleague, this compelling book explains the special theory of relativity to readers with no scientific background, using such familiar objects as trains, rulers, and clocks. 1960 ed. vi+72pp. 5⅜ x 8½. 0-486-42806-0

CATALOG OF DOVER BOOKS

A TREATISE ON ELECTRICITY AND MAGNETISM, James Clerk Maxwell. Important foundation work of modern physics. Brings to final form Maxwell's theory of electromagnetism and rigorously derives his general equations of field theory. 1,084pp. 5⅜ x 8½. Two-vol. set. Vol. I: 0-486-60636-8 Vol. II: 0-486-60637-6

QUANTUM MECHANICS: PRINCIPLES AND FORMALISM, Roy McWeeny. Graduate student-oriented volume develops subject as fundamental discipline, opening with review of origins of Schrödinger's equations and vector spaces. Focusing on main principles of quantum mechanics and their immediate consequences, it concludes with final generalizations covering alternative "languages" or representations. 1972 ed. 15 figures. xi+155pp. 5⅜ x 8½. 0-486-42829-X

INTRODUCTION TO QUANTUM MECHANICS With Applications to Chemistry, Linus Pauling & E. Bright Wilson, Jr. Classic undergraduate text by Nobel Prize winner applies quantum mechanics to chemical and physical problems. Numerous tables and figures enhance the text. Chapter bibliographies. Appendices. Index. 468pp. 5⅜ x 8½. 0-486-64871-0

METHODS OF THERMODYNAMICS, Howard Reiss. Outstanding text focuses on physical technique of thermodynamics, typical problem areas of understanding, and significance and use of thermodynamic potential. 1965 edition. 238pp. 5⅜ x 8½. 0-486-69445-3

THE ELECTROMAGNETIC FIELD, Albert Shadowitz. Comprehensive undergraduate text covers basics of electric and magnetic fields, builds up to electromagnetic theory. Also related topics, including relativity. Over 900 problems. 768pp. 5⅝ x 8¼. 0-486-65660-8

GREAT EXPERIMENTS IN PHYSICS: FIRSTHAND ACCOUNTS FROM GALILEO TO EINSTEIN, Morris H. Shamos (ed.). 25 crucial discoveries: Newton's laws of motion, Chadwick's study of the neutron, Hertz on electromagnetic waves, more. Original accounts clearly annotated. 370pp. 5⅜ x 8½. 0-486-25346-5

EINSTEIN'S LEGACY, Julian Schwinger. A Nobel Laureate relates fascinating story of Einstein and development of relativity theory in well-illustrated, nontechnical volume. Subjects include meaning of time, paradoxes of space travel, gravity and its effect on light, non-Euclidean geometry and curving of space-time, impact of radio astronomy and space-age discoveries, and more. 189 b/w illustrations. xiv+250pp. 8⅜ x 9¼. 0-486-41974-6

STATISTICAL PHYSICS, Gregory H. Wannier. Classic text combines thermodynamics, statistical mechanics and kinetic theory in one unified presentation of thermal physics. Problems with solutions. Bibliography. 532pp. 5⅜ x 8½. 0-486-65401-X

Paperbound unless otherwise indicated. Available at your book dealer, online at **www.doverpublications.com**, or by writing to Dept. GI, Dover Publications, Inc., 31 East 2nd Street, Mineola, NY 11501. For current price information or for free catalogues (please indicate field of interest), write to Dover Publications or log on to **www.doverpublications.com** and see every Dover book in print. Dover publishes more than 500 books each year on science, elementary and advanced mathematics, biology, music, art, literary history, social sciences, and other areas.